The Roman Auxiliary Units of Britain

Simon Turney

Pen & Sword
MILITARY

First published in Great Britain in 2024 by
Pen & Sword Military
An imprint of Pen & Sword Books Limited
Yorkshire – Philadelphia

ISBN 978 1 39905 603 8

A CIP catalogue record for this book is
available from the British Library

Typeset by Mac Style
Printed in the UK by CPI Group (UK) Ltd, Croydon, CR0 4YY.

MIX
Paper | Supporting responsible forestry
FSC® C013604
FSC www.fsc.org

Pen & Sword Books Limited incorporates the imprints of After
the Battle, Atlas, Archaeology, Aviation, Discovery, Family History,
Fiction, History, Maritime, Military, Military Classics, Politics,
Select, Transport, True Crime, Air World, Frontline Publishing, Leo
Cooper, Remember When, Seaforth Publishing, The Praetorian Press,
Wharncliffe Local History, Wharncliffe Transport, Wharncliffe True
Crime and White Owl.

For a complete list of Pen & Sword titles please contact:

PEN & SWORD BOOKS LIMITED
47 Church Street, Barnsley, South Yorkshire, S70 2AS, England
E-mail: enquiries@pen-and-sword.co.uk
Website: www.pen-and-sword.co.uk
or
PEN AND SWORD BOOKS
1950 Lawrence Road, Havertown, PA 19083, USA
E-mail: uspen-and-sword@casematepublishers.com
Website: www.penandswordbooks.com

Over the heather the wet wind blows,
I've lice in my tunic and a cold in my nose.

The rain comes pattering out of the sky,
I'm a Wall soldier, I don't know why.

The mist creeps over the hard grey stone,
My girl's in Tungria; I sleep alone.

Aulus goes hanging around her place,
I don't like his manners, I don't like his face.

Piso's a Christian, he worships a fish;
There'd be no kissing if he had his wish.

She gave me a ring but I diced it away;
I want my girl and I want my pay.

When I'm a veteran with only one eye
I shall do nothing but look at the sky.

(W. H. Auden, *Roman Wall Blues*)

Contents

Acknowledgements

The various academics whose works I have worn thin putting together this tome are all listed in the bibliography later, but certain people stand out for their aid and contributions. Most of the photographs in the book are my own, with a few in the public domain, but odd ones I have simply not had the opportunity to acquire were supplied by the excellent Pete Savin and Lesley Metcalfe, thanks to them both. The most knowledgeable man on the Roman military I ever met, and probably the most entertaining, Dr Mike (M.C.) Bishop never falters and never fails to throw possible answers at me when I stumble across questions. The fabulous Simon Elliott, an author of excellent Roman non-fiction himself, selflessly agreed to read a draft and comment, as did the amazing Graham Harris of Legio VI Victrix in York. Paul, Charlene and Cellan Harston of Deva Victrix have always been there to supply experimental answers to real questions, as well as excellent reenactment photos. The fab guys at Terra Dacica Aeterna were accommodating enough to kit up in full Dacian battle gear in insane heat just for me. Thanks, guys! My wonderful wife Tracey, my kids Marcus and Callie, and my parents Tony and Jenny, put up with a good two months of moaning as I tried to pull it all together towards the end, so kudos and thanks to them. And to everyone who has given me encouragement and claimed to be waiting impatiently for the release, you have only yourselves to blame!

I would like to dedicate this book to all great teachers and scholars everywhere, embodied for me by my grammar school history teacher, the late Dr Bill Petchey, without whose enthusiasm and passion I would have limped through life without knowing the meaning of the phrase '*terminus ante quem*.'

Introduction

There is, in the mind's eye of the British public, a certain image of the Roman soldier – an image formed in the Victorian era and earlier which lingers, which informed all early depictions of the Roman military, which formed the basis for modern Roman re-enactment, and which remains the 'Hollywood Roman'. That soldier wears a crimson tunic, the banded plate armour known to modern historians as '*lorica segmentata*', a helmet with a crest or plume, a rectangular shield emblazoned with lightning bolts, a short stabbing sword, and leather sandals.

Of course, our knowledge and understanding of the Roman military expands every year, with every new archaeological find and every new theory. That we have unearthed so many more remnants of scale armour, and whole new varieties of plate, are such discoveries. That the finds at places like Vindolanda now tell us that the wearing of enclosed leather boots was a lot more common than once thought, and that the military sandal was more or less phased out, are just examples of how our understanding is slowly changing that picture, struggling against two centuries of certainty in that traditional image.

But there is another angle that this ignores: that of the *auxilia*. The armies of Rome, in the form in which we tend to picture them, are the legions of the first and second centuries AD, not the Republican army from which they evolved, and not the later army which had adapted to face new enemies. And even in later days, those legions made up less than half of Rome's military. The rest was largely filled by the navy – the subject for *another* work – and the auxiliary forces, which Tacitus tells us were 'in not much inferior strength' to the legions.[1] There is a growing interest in this force that provided more than half of Rome's military might, and it is to those *auxilia*, in the framework of Britain, that this volume turns.

Many of Rome's military terms come down to us, whether from Latin or Greek, as familiar, and some have become part of our own tongue. 'Legion' has taken on a meaning beyond its origin, while the modern misuse of 'decimate' to mean the same as 'obliterate' sets my teeth on edge, but the survival of the

1. Tacitus, *Annals* 4.5.

term cannot be denied. It is in the nature of language that we look to the word to determine its meaning. In the term '*hippika gymnasia*', we do not need to look far, for the root of *hippo* tells us it is about horses, and *gymnasia* suggests sports and displays. That the *hippika gymnasia* was a cavalry sports event is clear. Similarly: 'gladiator', from 'one who uses a gladius.'

What, then, do we make of '*auxilia*?' In modern terms, auxiliary means a support, a backup, a helper, and that is exactly what the Latin term means. So we look to the origins of the *auxilia* and we can see that they were created as helpers, in this case to the legions that were the core of Rome's early army. This status as an 'aide' to the citizen soldiers is often visible in their history, especially in the earlier days of the Principate, and one does not have to look beyond Tacitus' account of the Second Battle of Bedriacum to see these 'helpers' living up to that name: 'Both generals had auxiliary infantry and cavalry, as well as fleets and allied kings.'[2]

This work began as an attempt to catalogue all the evidence for the auxiliary units that served in Britannia, and to attempt to build a service history for them, based around the framework of epigraphic and archaeological evidence, contemporary and later accounts, the geographical situation, and simple common sense. As such, those unit listings occupy the latter half of the book, and should give the reader any reference they require regarding any of the units in Roman Britain.

As this is a work purely on the *auxilia*, I have dealt with neither the Roman navy, nor the legions, excepting where such information has a bearing on the subject at hand. In the history of the *auxilia*, they follow something of an arc, from ally to standing army to barbarian levy, and so I have dealt with them from the time of the invasion in AD 43 until the close of Roman political and financial support of the province in AD 410.

It is my hope that this book will serve as a useful reference point for the scholar, as well as a general work on the island's auxiliary forces for the interested party. Where else to begin, then, but the beginning.

2. Tacitus, *Histories* 2.4.

Part I

Overview

Chapter 1

Historia

(A short history of the *auxilia* and their time in Britain)

From the early days of the Republic, in the sixth century BC, Rome developed a citizen militia, evolving from a military model based on the Greek phalanx, with innovations and influences from the various Italian peoples. The legions were born. In their earliest days, as later, the legions were divided into centuries, although in those times there may well have been 100 men in each century, which accounts for a term that actually had little relevance to unit strengths throughout most of Roman history. The legions evolved throughout the Republic, and though that evolution deserves deep study, this is not the place for it. By the mid-Republic they had moved away from a Greek model entirely and had started to become the force that would subjugate an empire. By the time of Polybius, the system had become a fierce and formidable fighting machine. In the later Republic, Gaius Marius would change the way the legions worked, dropping the manipular model and creating the 'Marius' Mules' that we now find so much more familiar. Caesar and Pompey would go on to change things again, a time of growth and invention for the army, and it was only under Augustus that things began to become settled fully. Indeed, until Augustus, Roman legions were not a standing army, but were raised for a campaign and then released once it was over.

Early Roman legions from the Altar of Domitius Ahenobarbus, c.122 BC, now in the Louvre. (*Photo: Wikipedia/Shakko*)

What of the *auxilia*, then?

Rome's army was formed of those legions, which were essentially a heavy infantry force. In the field against a similarly infantry-based force such as many they met in the early days of expansion, Rome could be confident in besting an opponent. Unfortunately, no army can rely so strongly on one type of troop and, as time passed, Rome became starkly aware of its failings.

In addition to the legions, Rome did have a cavalry force available. Naturally in a citizen army, just as the legions formed the free and enfranchised people of Rome into infantry units, the cavalry was also drawn from the citizen population, though there was something of an additional requirement. Cavalry needed horses, and so only citizens wealthy enough to own a horse could become cavalry. This, of course, meant that there was a distinctly limited supply of horsemen available to Rome, and the quality of such cavalry was based on financial superiority rather than martial skill, so the cavalry took on something of a secondary role in warfare.

Increasingly, Rome found itself facing peoples who had evolved over centuries into a primarily cavalry force. In 387 BC, she first met defeat at Gallic hands at the Battle of the Allia. Though we have only sketchy details of the campaign, Livy tells us of the Gauls that 'their horse and foot in an extended line covered a vast tract of ground.'[1] That the Gauls moved so swiftly they horrified the Romans suggests a strong mounted contingent, and indeed the warrior with whom Quintus Ambustus duels during the struggle is on horseback. By the time of Caesar's writings, he makes it clear how impressive and dangerous the Gallic and Germanic horsemen could be. Faced with such peoples, Rome's own citizen cavalry would have been woefully inadequate. Clearly the Romans would need to supplement their own forces with horsemen from outside to face such a threat. Throughout the time of the Republic, Rome would rely ever more heavily upon allied or mercenary horsemen, from Magna Graecia, and Gaul, Germania and Hispania. The most notable evidence of this is Caesar's *Gallic Wars* in which his effective diplomacy allowed him to field an army heavily supported by Gallic and Germanic cavalry against the enemy Gallic horse.

This situation of inadequacy did not just apply to cavalry, of course. Over centuries of expansion, Rome repeatedly came across enemies whose strengths only highlighted distinct weaknesses in Rome's own forces, whether that be tribes fighting guerrilla-style warfare in the mountains, expert archers or slingers, or even that greatest enemy of all: Carthage, with her impressive navy and war elephants. In the Punic Wars, the Romans quickly became aware of the inadequacy of their ships and were forced to adapt and expand in order to field sufficient power to face Carthage. Such, however, was the nature of Rome. When faced with such difficulties, it led the Romans to adapt, to create, to innovate and to change. Necessity is, as they say, the mother of invention.

1. Livy, *History* 5.37

Thus, while Rome could only field heavy infantry from its citizen army, with a lesser cavalry contingent, it became conversant in relying upon outside sources, or '*externi*'. Allies could supply troops for a campaign, and subjugated territories could be compelled to send forces. Where this was not possible, or was insufficient, there was always the possibility of employing mercenary groups. By hook or by crook, Rome was able to supply the forces it needed in any environment. Balearic slingers, Greek archers, Tarentine horsemen… Rome's heavy infantry would have the support it needed. Perhaps the epitome of this system is provided by the Third Macedonian War of 171 BC. Among the forces fielded by Rome for the campaign were Numidian cavalry and elephants supplied by Masinissa (an ally of Rome), 2,000 Ligurians, likely a forced levy as Liguria had been subjugated only a few years before, and 'a body of Cretan archers (whose number was not specified, the order only mentioning whatever succours the Cretans, on being applied to, should send)', which makes it clear these were a mercenary force.[2] As such, the force Publius Licinius commanded to be assembled included not only Rome's own army, but allies, levies and mercenaries.

This system of supplementing Rome's army with foreign forces grew more common and preferred as time went on, and by the end of the Republic, Sertorius's favouring of his Spaniards and Caesar's favouring of Germanic cavalry show how entrenched the system had become. Indeed, even into the Imperial era, this sort of favour continues to be shown, with Agricola's reliance upon Germans in his campaigns, and Trajan's use of Mauri cavalry in Dacia.

Such reliance, in fact, begins to shift the whole focus of the Roman military. As the efficacy of the system proves itself time and again, Rome pays less and less heed to the inadequate branches of its own military. The last direct mention we have of citizen cavalry among the legions is in Sallust's account of the Jugurthine War (111–105 BC), and though there may have been a cavalry contingent in some form over the following century, the reforms of Marius favour attaching auxiliary cavalry to the legions. Perhaps the most telling is that in the first year of Caesar's *Gallic Wars*, when he rides out to meet the German king Ariovistus he clearly does not have legionary cavalry available, and so for his escort he takes the best mounts from his native horsemen and gives them to men from Legio X (Tenth Legion) to ride with him, an escapade that earned that legion its name: X Equestris.

There is ample evidence that with this increasingly formalized system, the term '*auxilia*' was already in full use during the Republic. However, at this stage, those auxiliaries are, much like the legions, still temporary forces, raised or hired to provide manpower for a campaign, and then released afterwards. They

2. Livy, *History* 42.35.

Celtic warriors from the Gundestrup cauldron, first to third centuries BC, National Museum of Denmark. (*Photo: Claude Valette, CC BY-ND*)

are the ancestors of the forces this volume studies, but yet bear only a passing resemblance. And it is worth remembering throughout this book that the reason I include short-lived and irregular units is purely because the *auxilia* are 'helpers', and that implies a much less rigid and formal system than we often imagine.

The crucible in which the *auxilia* as we know them were born follows this period, in the last century BC, a period we know as the troubled late Republic and its fall. Throughout the preceding centuries, campaigns had been occasional and focused, but during the time marked by the rise of dictators such as Sulla, Marius, Sertorius and Caesar, Roman wars become something entirely different. Between the start of the Social War in 91 BC and the Battle of Actium in 31 BC, there were sixty years of almost constant conflict. When Rome was not at war with someone else, she was at war with herself.

91–89 – Social War
89–84 – First Mithridatic War
83–82 – Sulla's March on Rome
80–72 – Sertorian War
73–63 – Third Mithridatic War
62 – Rebellion of Lucius Sergius Catilina
61–60 – Caesar in Spain and Lusitania
59 – *a notable year of relative peace*
58–50 – Gallic wars
49–48 – Civil War
48–47 – Alexandrian War
47–46 – African War

45 – Spanish War
44–31 Caesar's assassination, civil wars

This impressive, and far from exhaustive, list makes one thing clear. For sixty years military units were raised and fielded, and while there were legions disbanded and allied troops returned to their lands as required, for six decades Rome played host to an immense force, spread across the Republic and engaged in one conflict or another. This, of course, came to a head in 31 BC, when Octavian and his officers met Antonius and Cleopatra in a naval engagement at Actium, off the coast of Greece.

For the first time in over half a century, Rome could hope for peace. And that was indeed immediately what the shrewd Augustus (still named Octavian at this time) set out to do, finally achieving full victory and returning to Rome in 29 BC. As the Republic was brought down from a state of constant war and agitation into a peace-time situation, its population could breathe once more. There was, however, one thing that rather stood out among the need for social and political reform, in the latter days of 29 BC. Rome was at peace, yet she was supporting a massive army.

Moreover, that massive army was no longer a force of temporary citizen soldiers, untrained levies or eager mercenaries. By this time, the various units at the fingertips of Rome's ruling body were grizzled, powerful veteran forces, be they legions raised by Caesar with twenty-eight years of service on their record, or Gallic cavalry forces that had moved from serving Caesar to serving his great nephew. What Rome suddenly had was a huge army of professional soldiers with no war to fight. Of course, by this stage, there had been war in the Republic for so long that no one alive would be able to remember a time when legions were levied for a campaign, and peace reigned thereafter. Rome's attitude to war had changed forever. If we are to believe the account of Cassius Dio, it was the statesman Maecenas who provided the answer for Augustus:

> A standing army also should be supported, drawn from the citizens, the subject nations, and the allies, its size in the several provinces being greater or less according as the necessities of the case demand; and these troops ought always to be under arms and to engage in the practice of warfare continually. They should have winter-quarters constructed for them at the most advantageous points, and should serve for a stated period, so that a portion of life may still be left for them between their retirement from service and old age.[3]

3. Cassius Dio, *Roman History* 52.27.

In short, the Republic had reached a point where the logical and, indeed, only *feasible* step was to legitimize and organize the huge force at its command into a standing army, just as Maecenas suggested. The units were there ready, the divisions between the legions and the *auxilia* in place, and very likely much of the administration and many of the conditions required would already be observed by those forces after decades of service.

We shall not deal in depth with the legions or the navy, but it would be remiss not to at least briefly mention them. We cannot estimate how many legions there were at the end of the civil war in 30 BC, but suffice it to say that between 58 BC and 44 BC Caesar had fielded at one time or another twenty-six different legions. Given that he was only one of several generals with an army at his command, we can suspect a staggering number of legions raised and often dismissed during that time. Regardless of the number of extant legions, Augustus gives us a figure for the actual manpower. In his Res Gestae, he tells us:

> The number of Roman citizens who bound themselves to me by military oath was about 500,000. Of these I settled in colonies or sent back into their own towns, after their term of service, something more than 300,000, and to all I assigned lands, or gave money as a reward for military service.[4]

More than 300,000 citizen soldiers retired! This means he kept in service between 100,000 and 200,000 legionaries. These were formally organized into 28 standing legions, and given an estimate of 5,000 for the manpower of a standard Augustan legion, that would total 140,000 men, which falls comfortably within Augustus's figures. So, more than sixty per cent of the citizen army was stood down and retired after the war, the rest consolidated in twenty-eight legions, which could then be assigned to troublesome areas of the Republic. It would be even more difficult to estimate fleet sizes, even given that Augustus goes on to tell us that he captured 600 ships of trireme size or greater.

What concerns us is, of course, the sudden enfranchisement of the auxiliary troops. Those who had been raised by levy from territories within the Republic could easily be instantly reorganized into regular units under the command of the Senate, attached to the legions and assigned to a region. One must wonder, though, how Augustus went about incorporating into his new standing army those units that had been present in the last days of civil war either as loaned troops from allies or as paid mercenaries. In the fanciful words of Cassius Dio, Maecenas had counselled the army should be 'drawn from the citizens, the subject nations, and the allies', but the question of *how* that was done remains.

4. *Res Gestae* 1.3.

The *Res Gestae of Augustus*, twentieth century copy in Rome. (*Author's collection*)

Presumably mercenary units could be given the option of a permanent contract as an auxiliary unit in the new system, or returning to their own lands. Having fought for years in Rome's campaigns they would be unlikely to refuse the offer. One might also picture Augustus writing letters to various allies and client kings and thanking them for their contributions, presuming that they will have no objection. After all, no one in their right mind would object.

So somehow, and much of the detail remains obscure, the non-citizen troops were drawn into the new standing army as a codified system of auxiliary units. Pay scales were secured and the term of service for an auxiliary soldier was set at twenty-five years, though that time could be extended for specific individuals of rank. We do not know of any retirement benefits agreed at the end of that term. Legionaries received their '*honesta missio*' upon completion of their term of service, with a gift of 12,000 *sestertii* from the treasury. Whether any lesser equivalent was part of an auxiliary's retirement package, we cannot say.

What we *can* say is that at some point between Augustus and Claudius, the one great benefit of retirement for an auxiliary was fixed. Upon completion of twenty-five years of service, that *peregrinus* (foreigner), was granted the coveted Roman citizenship that brought with it legal and social benefits. This grant of citizenship was likely an innovation of Augustus' as part of his grand reorganization, though the first confirmed evidence of it comes from the reign of Claudius in the form of a diploma. It may, alternatively, be that Claudius initiated the grant, or perhaps simply that before he began to issue diplomas there were no written records, or even that diplomas *were* issued earlier but none have survived.

If regular pay might not be enough of an incentive to join the *auxilia*, then the grant of citizenship was surely an immense draw. The recruits available for auxiliary units were free men, but not Roman citizens, and the protection of Roman law, the right to vote, and to enter the political world of the Republic

Diploma found at Lanchester, presumably of a soldier who retired from the fort there, now in Durham archaeological museum. (*Author's collection*)

(and then the Empire) would be tempting. Not only this, but upon retirement, citizenship was also conferred upon any extant spouse and children, as well as upon any future family. As such, completion of service was not just retirement, but a doorway to a newer, safer, more-civilized life for the whole family. For the *value* of that citizenship we need only look to a passage from the Bible:

> When the centurion heard this, he went to the commander and reported it. 'What are you going to do?' he asked. 'This man is a Roman citizen.'
>
> The commander went to Paul and asked, 'Tell me, are you a Roman citizen?'
>
> 'Yes, I am', he answered.
>
> Then the commander said, 'I had to pay a lot of money for my citizenship.'
>
> 'But I was born a citizen', Paul replied.
>
> Those who were about to interrogate him withdrew immediately. The commander himself was alarmed when he realized that he had put Paul, a Roman citizen, in chains.[5]

It might be noted that the grant of Roman citizenship to soldiers was not a new concept, a prime example being the Turma Salluitana, a Spanish cavalry unit who served in the Battle of Asculum in 89 BC and who were awarded en-bloc the coveted citizenship. However, previously such grants had been one-offs in return for specific services. Now, any *peregrinus* could become a Roman citizen after twenty-five years of service in the *auxilia*.

Our focus on Britain comes to the fore, of course, during the reign of Claudius. In AD 43, the army of Aulus Plautius set sail from Gesoriacum, after

5. Acts 22, 26–29.

a number of delays and problems, and landed without opposition at the future site of Richborough. This place, it might be noted, should be hailed as one of the most important sites of Roman Britain. Location of the original landing, main port access to the island from the rest of the empire, location of a grand triumphal arch, and one of the last surviving manned Roman installations on the island in AD 410, Richborough has seen more Roman history than almost anywhere else in Britain.

Plautius' force we know to have included four veteran legions (II Augusta, IX Hispana, XIV Gemina and XX Valeria), two of which would remain in Britain for centuries. Roman records tend to concentrate on the citizen troops, and so we have considerably less information on the auxiliary forces that accompanied the invasion. Plautius had come from Pannonia, bringing Legio IX with him, and it seems likely therefore that he brought auxiliary troops from the Danubian region also. What is clear is that he gathered forces from Germany before crossing, for we can be confident of a number of Batavian units present. Peddie estimates the full invasion force to have numbered between 40,000 and 50,000 men, not including naval and support units. Given a rough legion size of 5,000 men, and four legions present, that means we can expect between 20,000 and 30,000 auxiliaries to have been involved. A figure of 20,000 would be in line with Tacitus's suggestion that auxiliary forces roughly matched the legions in number across the empire, and would give us a force of 40 quingenary cohorts and *alae*, 10 cohorts to each legion. Given that we only know of 76 cohorts and *alae* that ever served in Britain, that's quite a figure, comparatively, at one time and in one force.

Of the makeup of the invasion's *auxilia* we can make only educated guesses. Based on surviving inscriptions and knowledge, we can suggest that six of our known *alae* were involved (Gallorum Indiana, I Hispanorum Asturum, Hispanorum Vettonum, Pannoniorum Sabinia, Pannoniorum Tampiana and I Thracum) and twenty-three of the cohorts (I Alpinorum, I Aquitanorum, I-VIII Batavorum, III Bracaraugustanorum, III Breucorum, I-IV Delmatarum, I-II Delmatarum Equitata, I-V Gallorum, I Hamiorum, VI and VII Thracum). Notably, given what I previously noted of Plautius's movements, of these twenty-nine units, fifteen are of Germanic or Gallic origin, and eleven of Pannonian or Danubian origin.

We are treated to a clear description of Batavian auxiliaries at the Battle of the Medway, for Cassius Dio tells us that Plautius 'sent across a detachment of Germans, who were accustomed to swim easily in full armour across the most turbulent streams.'[6] This is a known skill of the Batavians, referenced time and

6. Cassius Dio, *History* 60.20.

again in Roman history, and just thirty-four years later Agricola would employ them, probably the very same cohorts, in his invasion of Anglesey, as Tacitus tells us he: 'carefully picked out from his auxiliaries men who had experience of shallow waters and had been trained at home to swim carrying their arms and keeping their horses under control.'[7]

The story of Britain's conquest is an involved one, and occupies the bulk of the Principate, and I shall not detail the entire centuries-long campaign here, but to highlight pertinent moments such as the initial invasion of Claudius.

Empire-wide, the *auxilia* must have grown impressively throughout the first century, for most client states would have been able to field their own military, and upon the annexing of their lands by Rome, those forces could then be added to the ranks of the *auxilia*. Between 7 and 73 AD, Rome annexed, among others, the states of Cappadocia, Commagene, Mauretania and Thracia, all of whom would have been able to deploy their own troops, and all of whom are recorded as origins for the *auxilia*.

While having a ready supply of such forces was extremely valuable, the future also had to be taken into consideration. Augustus set out a system for an auxiliary term of service and retirement, and that meant there needed to be a system in place for further recruitment to replace those retirees. Initially recruitment was quite sensibly based upon the census. This would give a population listing for each province, which would allow the Senate and the consuls to judge a reasonable level of recruits for each region. Such matters would naturally be a balancing act between drawing the optimum number of soldiers without pushing the boundaries of reason and risking revolt. As we shall see throughout this work, despite being an empire of rigid systems, Rome was prepared to bend, break, or blatantly ignore any system that got in the way of efficient operation, and so we do not have to look too hard to find cases of individual treatment for a recruiting region.

Of those feared Batavi, who we have already met and will encounter several times in the coming pages, Tacitus tells us:

> They are not subjected to the indignity of tribute or ground down by the tax-gatherer. Free from imposts and special levies, and reserved for employment in battle, they are like weapons and armour – only to be used in war.[8]

7. Tacitus, *Agricola* 18.
8. Tacitus, *Germania* 29.

Thus we see how the Batavi were encouraged to supply large numbers of soldiers by way of being immune to taxes for the privilege. The Batavi do, indeed, produce those large numbers, supplying four known cohorts to Britain alone, though unit numbers reaching IX suggest there were at least five more. Despite the fact that it has been calculated that regions would generally supply less than one per cent of their manpower to the *auxilia*, Haynes suggests that almost every Batavian family would have a member serving in the Roman military!

The system in its early stages played to the familiarity of the units with their commanders and homelands. Cohorts and *alae* remained under the command of their own nobles, much as had been the case when they had been allies under Caesar in Gaul, or before. An early example of this is once more supplied by the Batavi, who were commanded by a chieftain of theirs called Chariovalda during Germanicus's campaigns across the Rhine. Indeed, such was the normality of auxiliary units being commanded by native nobles that many such early units took the name of that commander as part of their unit's appellation, as will become clear in the listings later in the book. The theory, clearly, was that a non-citizen unit led by their own would feel more confident in their commander, that they would have less to argue about. Moreover, those units would then be deployed in familiar home territory, once again just like Chariovalda's Batavi fighting for Germanicus around the Rhine. Not only does it involve less upheaval for the soldiers of the unit, it means there will be few barriers with language and culture, and they will be familiar enough with the terrain and the fighting styles and societies of the locals that the advantages are clear.

I daresay the reader can already see danger in such a system, though it is easily overstated. It should be noted that many of the lands in which such units are deployed were not kingdoms or empires such as Rome or Egypt, but lands that were at best loose confederations of tribes, often with little connection between them. Indeed, before the arrival of Rome, many of the Celtic tribes across northwestern Europe had spent much of their history fighting each other. That being the case, Rome might be quite justified in presuming it safe to field a native unit against their neighbours. In Britain alone we have one very clear instance of that situation. In AD 83, Agricola campaigned in Scotland and fought a confederation of Highland tribes at Mons Graupius, the culmination of the main campaign of conquest. Among the forces he fielded there, Tacitus tells us of: 'an army marching light, which he had reinforced with some of the bravest of the Britons who had proved their loyalty by long years of submission.'[9] Incidentally, it might be suggested that Rome's policies and activities with the

9. Tacitus, *Agricola* 29.

tribal systems of their neighbours were a catalyst in creating tribal coalitions that were then harder to defeat (the Caledonii, the Sarmatii etc.)

The idea of assigning *auxilia* to their home ground may have been general policy, but it was by no means absolute. In his *Annals*, our friend Tacitus tells us of the origins of the Thracian Revolt of AD 26 and states clearly that 'they refused to tolerate the military levies and to devote the whole of their able-bodied manhood to the Roman service.'[10] He goes on to explain the reason for this sudden refusal. They had always willingly accepted Roman service under their own chiefs because they were employed against neighbouring tribes, but rumour suggested they were to be posted among non-Thracians in distant lands. This is clear evidence that even as early as AD 26, auxiliary units were being posted far from home, and possibly, given the context, under non-Thracian commanders. This may, of course, be a limited example of fierce Thracians being used to bolster the numbers of other diminished cohorts, but it makes the point nevertheless.

The danger of units with local commanders on their home turf had been made clear more than once, in the Great Illyrian Revolt of AD 6–9, and then when Arminius of the Cherusci took advantage of his position as commander of an auxiliary unit to dupe Publius Quinctilius Varus into marching his column of three legions off into the German forests where they could be easily massacred. In AD 69, it happened again. Gaius Julius Civilis revolted in Germany and took a number of German auxiliary units with him. Cohorts that refused to rebel seem to have come away from the event with honours, and yet it took months to overcome Civilis' Batavi, the great but unpredictable general Petillius Cerialis finally putting an end to the uprising. Once Roman control was stamped upon Germany again, Cerialis took a number of units from the region and left to take up his posting as governor of Britain, transferring a swathe of Germanic troops to the troublesome island province, joining those already present. These Germanic units were to stay in Britain for centuries.

It is occasionally suggested that in the wake of the revolt, Vespasian changed the command system so that local officers would not command native units, but instead they would be placed under the command of equestrians – tribunes and prefects who were already citizens, and instead of serving a full term would be assigned for a matter of years before moving on to another post on the career ladder. The system of equestrian command, though, was set earlier than this by Claudius, as Suetonius makes clear: 'He rearranged the military career of the knights, assigning a division of cavalry after a cohort.'[11] Moreover, there

10. Tacitus, *Annals* 4.46.
11. Suetonius, *Life of Claudius* 25.1.

are examples of native commanders that date from after Vespasian's reign. Cohors VIIII Batavorum at Vindolanda was under the command of one Flavius Cerialis at the turn of the second century, a man who was very likely a Batavian himself, having gained his name from Vespasian for loyalty during the Civilis episode, and likewise the tombstone of a commander of III Batavorum (AE 1944, 97) notes that his wife was from Nijmegen, capital of Batavian territory, suggesting that he too was of Batavian origin.

In the wake of Civilis, there may have been a change to the general policy of recruitment, favouring the posting of auxiliary units far from their homeland to prevent their involvement in national risings, though we should consider this change a *trend*, rather than a *rule*. The garrisoning of Cohors I Germanorum in Germany, of XX Palmyrenorum at Dura Europos, of I and II Raetorum in Raetia, and of other such examples stand to prove that units could still be posted in their homeland, just as Flavius Cerialis proves they could still be commanded by their own countrymen.

From this moment we shall focus on Britain's *auxilia*, with peripheral details of the wider world. Britain's history through the Flavian era is characterized by a state of annual war and a constant expansion westwards and northwards until

Prefect Tiberius Flavius Miccalus, Istanbul Archaeological Museum (*Author's collection*)

the overcoming of the Caledonii in AD 83 and the conquest of Scotland. The campaigns and then the withdrawal of the army to the Tyne-Solway isthmus once more give us an era of constant movement and activity for the military on the island, which is reflected in the unit histories later in the volume. It has been suggested that following the murder and *damnatio* of Domitian, units around the empire hurriedly dropped any Flavian honorific in their name and replaced it with 'Augusta' to allow for the next regime without a connection to the fallen emperor. With *damnatio memoriae*, after all, the subject's name was removed from the public record.

During the brief reign of Nerva and then the reign of Trajan there is a time of relative peace and stability, and under Trajan we first see the introduction in epigraphy of something called a '*numerus*' (plural '*numeri*'). This term we will come across throughout the rest of Roman Britain and beyond, though the meaning of it is rather fluid, for the word can be more or less translated as 'unit'. From this period on in the second century we find more *numeri* appearing, applied to irregular groups of soldiers of varying number and form. They could be *numeri* of a certain type of troops, or from a certain tribe or region, those details showing in the name (Numerus Barcariorum, for instance, being a unit of boatmen, and Numerus Abulcorum seeming to reference the Abulci from Iberia.) The *numeri* were perhaps created on an ad hoc basis, echoing the Republican era's levies and hirelings, possibly for individual campaigns and duties that fell outside the purview of the regular army. *Numeri* of Britons are attested in the Odenwald by AD 145. It is likely that the *numeri* were referred to by this vague term simply because there was no standard to them as there was to the standing army. We cannot be sure of their organization, their status, their terms of service or even their pay. They remain a mysterious part of the Roman army which is constantly studied and reassessed.

Dedication from the House of the Vestals in Rome, with Domitian's name removed, in situ, Rome. (*Author's collection*)

In the second and third centuries these *numeri* generally seem to be commanded by a '*praepositus*' (plural '*praepositi*'), a similarly vague and fluid word that does not denote a specific rank, but rather a position of temporary command that could be held by any rank. Often, the role of *praepositus* seems to have been filled by legionary centurions. By the time of the *Notitia Dignitatum*, in the early fifth century, they are noted as under the command of prefects and tribunes, which has led to the suggestion that at some unknown point these *numeri* were legitimized and brought into Rome's standing army as a regular type of unit. As with almost every aspect of the *numeri*, this continues to be debated.

Similarly, the term '*cuneus*' (plural '*cunei*') starts to come into use at this time. It might be best to think of the word *cuneus* as 'formation', which allows just the same sort of vague application as we have with *numerus*. There are easily refuted suggestions that a *cuneus* was the mounted equivalent of the *numerus*. Possibly the two words are interchangeable.

One theory of *numeri* and *cunei* is that they occupied the smaller military posts around the empire, including fortlets, signal stations, watchtowers, milecastles and the like. Though this cannot be proven, it does make sense, for irregular units with the skills and familiarity required for certain duties would be perfect for some such sites. *Numeri* and *cunei* begin to be attested in Britain by the mid-third century, and appear in the record from then on.

The *auxilia* undergo something of a renaissance in the reign of Hadrian. Thus far they had been part of an army of expansion, their primary duties alongside the legions, conquering lands, pacifying and Romanizing them before moving ever onward. It is natural to suggest that the whole role of the *auxilia*, and indeed the army as a whole, began to change with Hadrian. In his reign, for the first time, permanent borders were created and defensive systems constructed along them. It is notable that from that reign, the three legions in Britain become settled after a century of regularly moving. Legio VI remain in York, and XX in Chester, until the fifth century. Legio II remain at Caerleon for some time, moving to Richborough by the fifth century. With the creation of Hadrian's Wall (and similar defensive systems in other regions) it seems only natural that the units moved into the new forts were intended as a border force. The focus moves from expansion to defence, and for the first time, the *auxilia* become the standing garrisons of Rome's frontier, as well as occupying hinterland forts where military control remained important.

The following half century saw another change, as Antoninus Pius came to the throne and decided to advance the frontier to the Forth-Clyde line. Alongside the legions, large numbers of auxiliary units were moved north and involved in both the construction and garrisoning of this new frontier. The occupation of the Antonine Wall was only to last for twenty years before the army returned to the previous frontier and reoccupied Hadrian's Wall, around AD 165.

From the reign of Antoninus (around AD 140) the conditions on a diploma granted at the end of a soldier's term of service changed. The grant of citizenship would now apply to the discharged soldier and to any future children, but no longer applied to *existing* children. What the motivation was behind this move we do not know, and it cannot have been a popular one among the potential recruitment pool. Perhaps this helps explain the huge sum the next emperor, Marcus Aurelius, paid the army on his accession.

From the time of Marcus Aurelius through both his reign and that of Commodus, the military in Britain enjoyed a time of relative stability and peace. Forts were reconstructed and building work is widely attested, but for thirty years there was no great campaigning activity in Britain, the imperial military focus being solidly set on the Danubian border and the Marcomannic Wars.

The next phase of major activity for the *auxilia* comes with the rise of Septimius Severus. From AD 208 to 211, Severus and his sons, as well as a large field army, arrived in Britain with a view to repeating Agricola's successes and completing the conquest of the island. The *auxilia* were once more moved north

The Marcomannic Wars as depicted upon the column of Marcus Aurelius, Rome. (*Author's collection*)

en masse. Hadrian's frontier system was reworked, in particular the fort at South Shields, which was extended and refitted to act as a supply base with numerous granaries to support northern activity. The Antonine Wall was reoccupied, and both legions and *auxilia* would move into Scotland and campaign for several seasons. Severus' death in AD 211 cut short this campaign, and Caracalla abandoned the notion of adding Scotland to the empire, yet again pulling the border back to Hadrian's Wall. Notably, during Severus' stay on the island, the province was split in two, becoming Britannia Superior and Britannia Inferior.

The accession of Caracalla was to have another effect on the *auxilia*. In AD 212 the Antonine Constitution granted Roman Citizenship to all free people across the empire. From that moment, the social distinction between legionary and auxiliary was largely removed, since the *auxilia* were now Roman citizens. Of course, this means that there was no longer a great need for citizenship to be granted upon retirement, and that is likely why we see no evidence of military diplomas after the reign of Severus. A question that remains to be answered is what effect this had upon recruitment into the *auxilia*. Previously one of the great draws to sign up was the promise of citizenship, but that lure had now vanished. The problem is that the half century which followed saw a series of barbarian incursions, wars, civil wars, plagues and other disasters, referred to by historians as the 'Crisis of the Third Century', and the drop in the number of auxiliaries empire-wide is noticeable, estimated at around a third of all manpower. How much of that can be put down to the lower incentive to sign up is impossible to tell, given all the other effects. It is worth noting an increasing reliance upon recruitment from barbarian populations beyond the empire's bounds into the *auxilia* from this time, for they would still yearn for a grant of citizenship. Caracalla's large increase in military pay (see later for details) probably softened the blow a little.

In Britain, we see a steady decline in the number of troops throughout the remaining centuries. In addition to war, plague, and changes in recruitment, a series of important figures would from this point begin to strip the military from Britain for service overseas. In AD 260, Gallienus took vexillations (temporary detachments) of two legions for service in Pannonia, Constantius removed troops in AD 306, Magnentius in AD 350, Magnus Maximus in AD 383, and Constantine III in AD 407. We do not have specific figures for these removals, though we can speculate that the *auxilia* took the brunt of it.

In the third century a new frontier began to rise to the fore in Britain. Hadrian's Wall was still occupied, but there seems to have been little threat to security in the north, and the garrisons were probably considerably reduced in size by this time. The wall and its forts even began to fall into decay. The focus had begun to shift to what is now termed 'the Saxon Shore'. A new system of forts began to arise to ward against invasion and piracy from across the sea. The dating for the Saxon Shore forts continues to be debated and reassessed, but it seems that the new forts of Caister-on-Sea, Reculver and Brancaster were all constructed around AD 220, and by AD 280–290 the entire system of eleven forts, stretching from the Wash in Norfolk to the Solent in Hampshire were in place, later to be extended with further coastal forts, fortlets and signal stations.

If the Antonine Constitution had begun the process of removing the divide between legions and *auxilia*, that divide was clearly fading rapidly by the time of these forts. Legio II Augusta is last attested in its home fortress of Caerleon around AD 259, and the legion is noted as the garrison of Richborough shore fort in the *Notitia Dignitatum*, suggesting that perhaps they were the builders of that fort sometime just after AD 273. Brancaster was occupied by Cohors I Aquitanorum from around AD 220, and by the time of the *Notitia* it had been replaced by a Dalmatian cavalry unit, while Reculver seems to have been occupied by Cohors I Baetasiorum from AD 220 until the end. Thus we can see there seems to be little difference in status between auxiliary cohort and legion, given that they are performing the same duty in adjacent forts by the late third century.

That century brought huge upheaval for Britain. In AD 260, spurred on by political instability and repeated barbarian incursions, Postumus, governor of Germania, seceded from the empire and formed a new 'Gallic Empire', which controlled the territories of Germany, Gaul, Iberia and Britain. Though ultimately it was to be retaken by Rome, the Gallic Empire survived as an independent state for fourteen years. During this time, much of the work on the Saxon Shore was begun, and the military of Britain must have been reorganized and moved, with Legio II Augusta being made part of the new system.

We cannot say with any certainty what consequences the secession from Rome had on the *auxilia* in Britain, but one particular effect leaps to mind. Recruitment and retirement would probably change little, since by this time the cohorts were undoubtedly largely manned by men recruited locally, bolstered periodically by groups from overseas. Command of a cohort or *ala* would have to change, though. Commanders of an auxiliary unit, be they tribune or prefect, were drawn from the central provinces and appointed by the Senate for a term of two or three years. Suddenly there was no connection to the Roman Senate, the commanders now belonged to a new empire, and there would be no replacements sent by Rome. How this was overcome we cannot be sure, though it seems likely the breakaway emperors would have had their own senate and consuls on the Roman model and would take on the appointment of commanders from the capital, probably at Trier or Cologne.

The Gallic Empire was not the only breakaway state, either. In the east, only a decade later, Palmyra suddenly took huge territories away from Rome.

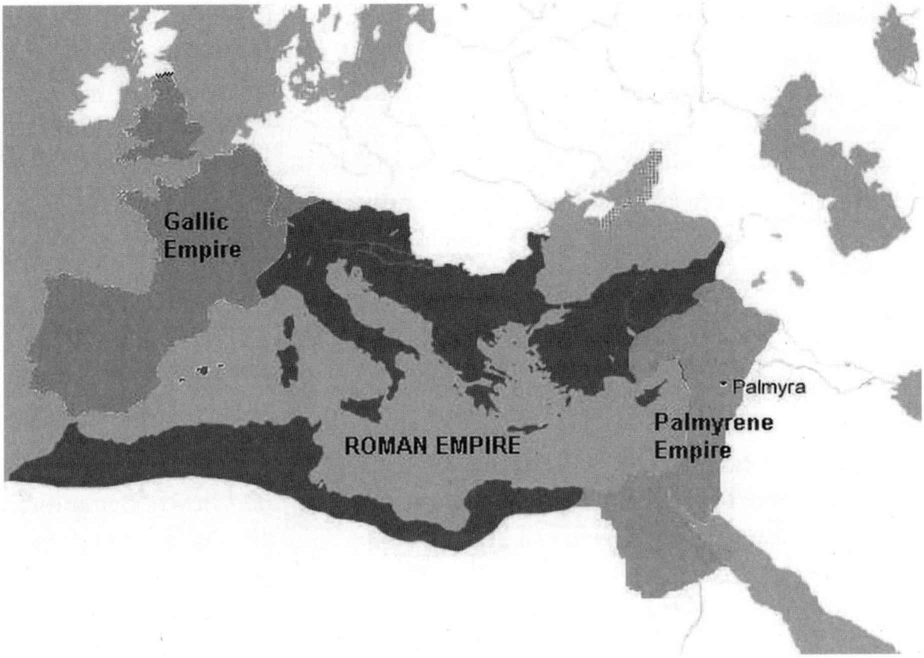

Breakaway empires c. AD 270. (*Photo: Wikipedia/Justinian43*)

The Palmyrene Empire was retaken and reintegrated with Roman territory by Aurelian in AD 273, and the Gallic Empire hot on its heels in 274, and it is possible that the emperor brought a number of auxiliary units to the island after this (the so-called Equites Scutarii Aureliaci of the *Notitia Dignitatum* spring to mind). He certainly engaged in a reorganization of Hadrian's Wall, and repairs and alterations date to this period.

With the accession of Diocletian in AD 284, we can more or less put the final nail in the coffin of the differences between legion and *auxilia*. Under the Tetrarchy, the empire's entire military system was reformed. The new focus is known as 'defence in depth'. The idea is that rather than attempt to secure the empire's borders with sufficient forces to stop incursions, which had proved impossible anyway for a century or more, it would be better to man the border with units who could act as scouts and watchmen as much as guards and soldiers, with the concentrated might of the military held back from the border as a 'mobile field army' that could be committed to any trouble spot with speed and efficiency. This policy we see increase and take effect more and more in Britain over the following century and a quarter.

Given the increasing focus on cavalry, and the need for the field army to be swift, it is possible that many of the existing legions around the empire supplied troops for the field armies, the *comitatenses*, as would auxiliary cavalry and infantry.

We know that the legions continued to exist following the changes, though their size is estimated to have decreased from around 5,000 to just 1,000 under the Dominate. That the legions in Britain continued to remain in garrison, but were now of less importance than a field army made up of various forms of troop, puts them in more or less the same situation as the *auxilia*. Together, they have become the other force of the late Roman military: the *limitanei*, with duties to oversee the borders.

There could be an argument for ending any study of the *auxilia* with the changes in the Diocletianic period, but since those units seem to have continued to serve, and new, different units may not be officially *auxilia*, but fulfil that same ancient role of 'helpers', it seems appropriate to continue our study for the following century.

A new upheaval was to follow on swiftly. In AD 286, a powerful soldier by the name of Carausius, commander of the fleet, the Classis Britannica, once more seceded from the empire, taking with him Britain and the coastal region of northern Gaul. We have sparse information on the island's military during this period, though it seems likely that little change occurred and that the trends already in place continued. The Gallic regions of this breakaway empire were retaken for Rome by Constantius Chlorus in AD 293, and Britain was to follow in AD 296.

Along with the changes in the military now came changes in the provinces in which they served. As Diocletian split the provinces into smaller units, so Britain was once more divided. We cannot be certain of the precise geography of these provinces, but the map below is based upon likely options. Britannia Prima in the west with a capital at Cirencester, Britannia Secunda in the north with one at York, Maxima Caesariensis in the south with London, and Flavia

The provinces of Late Roman Britain. (*Author's own work*)

Caesariensis in the middle with Lincoln. For the northern land 'Valentia', see below.

Two new officials would command the island's military: the *comes littoris Saxonici per Britanniam* overseeing the Saxon shore, based perhaps at Canterbury, and the *dux Britanniarum*, with command of the north of Britain, probably based at York. There is no surviving evidence for the mobile field army of Britain, and it may be that it was drawn of specific vexillations as and when required. This is just one of many unanswerable questions of late Roman Britain.

It is possible that one of the greatest indicators of the whole 'defence in depth' system at work is visible in *RIB* 653 from York. This altar was raised by one Marcus Minucius Audens, who is described in the inscription as a soldier and a *gubernator* of Legio VI. A *gubernator* was a ship's pilot. Clearly Legio VI in York had some sort of naval connection, presumably via the rivers Ouse and Humber. That York is generally assumed to be the central control site for the late Roman system of signal stations on the Yorkshire coast, and that the city was a provincial capital and home to one of the most important military officials of the era, shows how the system could work.

The world of the *auxilia* was changing, and perhaps more so in Britain than in many other provinces. The distinction between the old-fashioned legion and the auxiliary had more or less gone, in its place a distinction between the powerful central field armies and the *limitanei* who would occupy postings on the borders of the empire.

Britain was not to enjoy a rest. Just a year after the recovery of a second breakaway Britain, we find the first ever reference to a people of northern Britain by the name of the 'Picts'. This new tribal group, perhaps formed largely of the Caledonii, seem to have attacked Roman sites in northern Britain some time in the following years, for there are signs of destruction at the outpost forts of Hadrian's Wall including High Rochester and Risingham. They may have come further south, attacking and even crossing the wall, for something drew the attention of Constantius, now senior emperor of the west. The emperor arrived in Britain in AD 305 with an army and immediately went on the attack, campaigning in the north against the Picts. Constantius celebrated a victory, but died soon after, his son Constantine being raised to the purple at York in AD 306. The army in Britain likely felt a distinct attachment to the Constantinian dynasty. After forty years of breaking away from Rome only to be retaken, they could enjoy a feeling of security, had celebrated a victory in the north in the presence of an emperor, and had played host to the imperial court, even raising Constantine to the throne by the power of military acclamation. When Constantine returned across the Channel to campaign against Maxentius and Galerius, he almost certainly took a number of British units with him to supplement his forces.

There follows a series of works in Britain, beginning in AD 306, while Constantine was still here, and then continuing when he returned to the island once his throne was secure. Hadrian's Wall was repaired once more, along with alterations to the forts. A number of organizational changes were made and there is evidence of reduced garrisons in the reorganization of barrack blocks. Another possible result of this Constantinian reworking is the abandonment of all sites north of the wall. Risingham, High Rochester, Bewcastle and Netherby, the wall's so-called 'outpost forts', show no sign of military occupation after this time, and odd coin finds might as easily be linked to locals occupying the sites after the military have left.

Emperor Constans visited Britain in AD 343, following a campaign against the Franks. We cannot be certain of the reason for his visit, but it is entirely possible that the tribes were on the move once again in the north. It is likely that during his visit Constans made treaties with the Picts, and with the Scotti of Ireland, for just seventeen years later, in the reign of his cousin Julian, we are told by Ammianus Marcellinus (Ammian):

> in Britain in the tenth consulship of Constantius and the third of Julian raids of the savage tribes of the Scots and the Picts, who had broken the peace that had been agreed upon, were laying waste the regions near the frontiers.[12]

The tribes were supposedly at peace, then. In AD 350, as earlier noted, the would-be usurper Magnentius made his bid for power and in the process drew a number of the units from Britain overseas with him. A period of tribal peace would certainly make such a move less apparently dangerous.

When the Scotti and the Picts rose up in AD 360 as noted above, Julian sent his general Flavius Lupicinus to Britain to put down the rising. We have no information of the general's campaign in Britain, and he was back across the Channel swiftly, where he was arrested in the opening days of the civil war between Julian and his cousin Constantius II. Whatever Lupicinus did in his brief time on the island, it had little effect, for there were further raids in AD 365, and then in 367 the 'Great Conspiracy'.

According to Ammian, the garrisons of Hadrian's Wall, disenchanted with their masters and of an increasingly non-Roman nature, along with the '*miles Areani*', whose job it was to report on Barbarian trouble, sided with those

12. Ammian, *Roman History* 20.1.1.

Building inscription from Ravenscar signal station, possibly naming a *magister turris* called Justinianus, now in Whitby Museum. (*Author's Collection*)

barbarians and allowed the Picts to cross the wall into Britain. At the same time, the island was hit by forces of Scotti from Ireland and Saxons and Franks from across the Channel. The damage and chaos must have been tremendous. Valentinian I sent three men in quick succession to deal with the problem, withdrawing first Severus and then Jovinus swiftly, before handing the task to Flavius Theodosius. The general arrived at Richborough with a force of 2,000 men formed from the Heruli, the Jovii, the Batavi, and the Victores.

Theodosius was clearly equal to the task, and within the year had retaken control of the island, driving the barbarians back either north of the wall or to their ships. The traitorous *miles Areani* were disbanded. There followed another period of reorganization and building works in Britain. The wall was changed once again, and now it would appear that there was a decline in the settlements surrounding forts, some of which began to be abandoned around this time. With a smaller garrison and a rearrangement of the fort's interior, the civilian population outside could be brought safely inside the walls, and it is likely at this time that soldiers' families began to live with them in the barracks. It may also be at this time that the line of signal stations along the Yorkshire coast was created.

Two late Roman sources mention a fifth British province by the name of 'Valentia', with Ammian telling us that Count Theodosius in AD 369:

> restored cities and fortresses, as we have already mentioned, and established stations and outposts on our frontiers; and he so completely recovered the province which had yielded subjection to the enemy, that through his agency it was again brought under the authority of its legitimate ruler, and from that time forth was called Valentia, by desire of the emperor, as a memorial of his success.[13]

The suggestion that the land retaken consisted of the territory between the Hadrianic and Antonine Walls is an attractive one, and if this was the case, a natural capital for it would be Newstead (Trimontium), which had played a crucial role in all activity north of the Tyne-Solway line for three centuries.

Another decade or so of a breather, and Britain once more headed for trouble. The rebel Magnus Maximus made his bid for the throne in AD 383, and stripped the island of yet more of its military. It may well be at this time that the units named on the *Notitia Dignitatum*, the Milites Andereiani and the Seguntienses, are moved to their foreign locations, retaining the name of the garrisons they left (Andereiana undoubtedly being Anderida/Pevensey and Seguntienses being Segontium/Caernarvon).

We are left with an inescapable conclusion. By the close of the fourth century, repeated incursions, combined with repeated withdrawal of troops across the Channel, must have left Britain with a woefully inadequate defence system. We know of no specific mobile field army on the island, and if there had been a standing one for a century, undoubtedly one of the many failed usurpers would have taken them away by now anyway.

There is a strong suggestion that the great general Flavius Stilicho, father-in-law of Emperor Honorius, visited Britain in AD 395, and reorganized the military one final time. If it was not Theodosius who created the Yorkshire signal stations around AD 370, then they may well have been the work of Stilicho in 395. It may be that, knowing the danger to the empire had never been as great as it currently was, he was assigned to make Britain safe. With too few troops, the only way that could happen would be to rely upon naval support, signal stations and small but mobile units, further enhancing the 'defence in depth' system. Even if that might have staved off what was now probably inevitable, the last nail in the coffin was to come in AD 407 when Constantine III, yet another usurper, took more troops from the island in his failed bid for power.

13. Ammian, *Roman History* 28.3.1.

Fortlets and signal stations shown on Trajan's Column, Rome. (*Author's Collection*)

A map of the coastal defences of Britain by AD 410. (*Author's work*)

The *Notitia Dignitatum*, which probably dates to only a few decades later, tells us that the troops garrisoning Hadrian's Wall were still cohorts and *alae*, commanded by prefects and tribunes, that the interior forts of the north were occupied by legions or by *numeri* and *equites*, all under the command of prefects, and that the Saxon shore held a mix of all these units, generally commanded by *praepositi*. This is our last evidence for the Roman army in Britain. In AD 410 came the famous '*Rescript' of Honorius*. There will ever be arguments concerning this infamous missive, but Occam's Razor suggests that the letter was exactly what has been handed down to us by history.

Zosimus tells us that 'Honorius, having sent letters to the cities of Britain, counselling them to be watchful of their own security…'[14] and Gildas echoes this with:

> The Romans, therefore, left the country, giving notice that they could no longer be harassed by such laborious expeditions […] but that the islanders, inuring themselves to warlike weapons, and bravely fighting, should valiantly protect their country.[15]

This would not be enough. Britain would be hit again and again by raids in the ensuing century, and then by invasion after invasion. There are hints of Roman involvement in the island after AD 410, despite the *Rescript*, and the *Notitia Dignitatum* itself may be a clue as to plans for the reconquest of Britain, but the island was never to be recovered. We have no details on what military remained in Britain in AD 410, other than the *Notitia*, and even if that can be trusted (more on this later), it gives us no figures to identify strength. It seems likely that only small, largely native, garrisons remained, totally inadequate to defend the island. They may have turned into mercenary forces, lacking funding from Rome, and perpetual attempts to link them to the legend of King Arthur illustrate how, even now, the legacy of Britain's *auxilia* still fascinates us.

14. Zosimus, *New History* 6.
15. Gildas, *De Excidio Britanniae* 2.18.

Chapter 2

Ordinatio

(The organization of the *auxilia*)

The first thing to note with research into the Roman *auxilia* is how little actual information we have to go on. There are snippets and hints spread across dozens of ancient works, and archaeology provides sporadic proof, but common sense has to fill in the gaps until more concrete evidence is unearthed.

We know that from the outset *auxilia* were attached to legions (literally as 'helpers'). Tacitus, writing about the troubles in Germany, tells us of: 'eight cohorts of Batavians, auxiliaries belonging to the Fourteenth legion',[1] and a number of examples exist of the phrase 'legion … and its auxiliaries' or similar. In the first century, following Augustus' formation of the standing *auxilia*, there seems to have been a system of command in which the *auxilia* based within a certain catchment area reported to the legion at the heart of the zone. This makes a certain amount of sense. While the commanders of legions would naturally report directly to the governor, there would be only a handful of such men (three for most of the history of Roman Britain), while the number of auxiliary cohorts and *alae* proliferated. I suspect having in excess of fifty auxiliary prefects and tribunes reporting to him as well might have had the governor tearing his hair out. If, however, those men were divided up among the three legions, we end up with a hierarchy, something that Rome did very well. This situation of the *auxilia* reporting to the legions seems to peter out at the start of the second century, and it is tempting to see that as a reaction to the change in the entire nature of the army under Hadrian, when the *auxilia* shift from being an active part of a force of conquest to being border control garrisons. As such there would be considerably less need for ties to the legions.

We have already encountered such difficulties in attempting to estimate the size of the auxiliary contingent gathered for the invasion of Britain. Knowing that, it should be no surprise that estimating the size of Rome's entire standing auxiliary force is even more difficult. Cheesman, who remains the modern

1. Tacitus, *Histories* 1.59.

world's go-to source for auxiliary matters, goes into great detail, building a figure based on diplomas and upon the works of Josephus and Tacitus. From Vespasian's Judean armies, Vitellius's forces moving against Rome, a figure from a Mauretanian diploma, and Tacitus's statements of manpower, we can draw a figure of 130,000 covering the East, the Danube, Germany and Raetia, and Mauretania in the Flavian era. Cheesman estimates another 50,000 for the remaining provinces, bringing us to a total of 180,000 auxiliaries across the empire. He also estimates that figure rising over time, passing 220,000 by the mid-second century, based on the known increase in legionary figures over that time. With what we know of unit sizes, this gives us a number in excess of 300 cohorts and *alae* across the world, rising as time passes, a huge number of units and men. Webster suggests a number of auxiliaries ready for discharge annually in excess of 5,000 men.

The distribution of the army about the provinces naturally shows a concentration of strength on the borders and in troublesome areas, with only small numbers deployed in the hinterlands for various duties of control. Based on the evidence of Spaul and Holder, from the known cohorts and *alae* (to a total of 379 units) we can build a list of the Hadrianic deployment locations:

Border provinces		Hinterland provinces	
Britain	59	Dalmatia	3
Cappadocia	18	Iberia	7
Dacia	48	Thracia	2
Egypt	17	Macedonia	1
Germany	47	Sardinia	1?
Moesia	29	Anatolia	2
North Africa	40		
Pannonia	33		
Raetia and Noricum	21		
Syria and Arabia	51		

The concentration on the borders is clear. The slight trend towards strength in Iberia compared with other hinterland provinces likely points to its importance as a source of gold and silver, the control of which would require military strength. Other sites of such wealth production (Romania, Wales, Egypt etc.), as border provinces, were already garrisoned heavily. The duties of the *auxilia* in non-frontier regions would presumably fall into the category of 'policing'.

Our knowledge of the deployment, recruitment and retirement of auxiliary units and their commanders comes largely from epigraphy and military diplomas.

We have grazed the subject of diplomas already, but their value in research calls for a little explanation. From at least the reign of Claudius until at least that of Caracalla (the earliest known being AD 52, the latest AD 203), a retiring auxiliary soldier would receive a hinged bronze tablet that stood as proof of his service and consequent citizenship and rights. These documents, many of which survive in fragments, some in full, give us a great deal of information. The inclusion of the emperor's name and full titles, and of the province's governor, can usually give a close date for the document, even down to the month. Each diploma names the unit in which the soldier served, but also names the province, as well as a list of all the units being retired from that province, which gives us details of what units were in what province at what date, one of the most valuable resources we have. Of course, they also give us the name of the soldier, and any pertinent information about rank and position. Furthermore, they name his commander, therefore giving us the name of the unit's prefect or tribune in that province at that date. These tablets were kept by the soldier and appear all over the empire, for the retired soldier would not necessarily stay in the province in which he served, let alone the specific location, but copies were also kept in Rome. Diplomas fade from usage in the Severan era, likely due to Caracalla's widespread grant of citizenship, which would more or less nullify the point of a diploma.

In terms of stone inscriptions, we have several types of source. Tombstones at forts can give us a soldier's name and unit, and sometimes his origin and

The Edict of Caracalla, papyrus copy in Giessen University. (*Image courtesy of Leipzig University Papyrus Portal*)

other lesser detail. Building inscriptions and altars regularly name the unit, and often supply the names of individuals, though these are usually the commanders rather than the ordinary soldiers. We can also gain information more obliquely from inscriptions in other countries. Often, men of rank would vote statues, dedicate buildings, or simply pass away, and on the appropriate stonework, their career would be listed. This can be invaluable, as often he will have served as a prefect or tribune of an auxiliary unit in Britain at some point in that career. Other miscellaneous written sources exist to throw light on the world of the *auxilia*, such as the Vindolanda tablets, which give us much more day-to-day, prosaic information about the units and their staff.

With the knowledge now of estimated numbers, of distribution, and of the available sources of information, the next question is one of the creation and manning of auxiliary units. During the early Principate, auxiliary units were recruited, as noted earlier, based on a system drawn from the census. Rome, ever efficient, gathered types of units from those peoples famed for certain skills (archers from Syria, horsemen from Spain, Gaul and Africa for example).

There was, needless-to-say, a lean towards drawing the *auxilia* from the more peripheral provinces. The senatorial provinces, settled and long-since civilized, produced relatively few recruits. These provinces had acquired more and more grants of citizenship as time went on, leaving fewer and fewer individuals available for auxiliary recruitment. As such, it fell increasingly to the regions towards the edge of empire to provide the *auxilia*'s manpower. Again, we have already discussed how early recruitment shifted from local to more distant sources, but I shall state the case once more. With the empire's borders being the place such manpower was needed, it was the early standard to recruit units from local tribes to serve in that province. This was, however, by no means a rule set in concrete. A number of object lessons in betrayal changed Rome's attitude and, from the latter first century onwards, troops were more often sent elsewhere to serve, though again there is plenty of evidence to suggest that this was not an absolute rule.

We have already seen how, in addition to the enticement of regular pay and a grant of citizenship on retirement, other incentives could come into play in recruitment. That the Batavi were exempt from taxes on the basis that they supplied a large number of men to the *auxilia* shows how flexible the Roman systems could be. The only thing Rome was better at than creating efficient and rigid systems was finding ways to bend, break and circumvent those systems as circumstances required.

Vegetius remains our best ancient source for details of the military, even given the lateness of his writings. The problem, as usual, is that he writes about the *comitatenses* and the field army, not the *limitanei* and *auxilia*, and so for all the depth of his information, it is only of use to us as suggestions. With regard to the legions, Vegetius tells us that for recruitment age, puberty was the early standard, and Hadrian is noted in the *Historia Augusta* as setting the minimum age at the point the recruit's 'strength warranted'.[2] This is somewhat supported by epigraphy, with such individuals as Marcus Caesius Verus noted on his tombstone (*AE* 1990, 896) as dying at 41 after twenty-seven years of military service, putting his recruitment at the age of 14. This would see him just having taken the *toga virilis*, making him officially a man. The upper levels of recruitment age seem to have petered out by the mid-30s, with examples such as Marcus Flavius Flavianus (*AE* 1908, 226), who died at 52 after twenty-two years of service, making him a thirty-year-old recruit.

Dacian Archer as portrayed by a reenactor of Terra Dacica Aeterna. (*Author's collection*)

Given this wide variation even among the legions, which were surely more standardized, we can only guess at how wide recruitment selections were among the *auxilia*. Given that in early auxiliary recruitment it seems that units maintained very much their regional flavour, it may be that each unit was recruited according to entirely different criteria that were appropriate to the local populace. Certainly we can see unique recruitments happening from time to time. A case in point occurred in AD 175. We are told of Marcus Aurelius and the Iazyges (a Sarmatian tribe) in the Marcomannic Wars:

> The Iazyges were defeated and came to terms [...] and that they promptly furnished as their contribution to the alliance eight thousand cavalry, fifty-five hundred of whom he sent to Britain.[3]

The Iazyges then, who we will meet in the unit listings as the Ala Sarmatarum in Ribchester, were recruited en masse as tribute from a beaten foe, and would have been incorporated into the army in very much their tribal state.

In the early days of the standing *auxilia*, units were generally drawn from a specific area, often recorded in the unit's name. We can be fairly certain of the

2. *Historia Augusta*, Hadrian 10.8.
3. Cassius Dio, *Roman History* 72.16.

ethnic makeup, for instance, of Cohors I Aquitanorum when they were initially created, just as we can be of Cohors V Raetorum, or the Ala Sarmatarum. What becomes clear is that as time goes on there is a natural general shift from local to open recruitment. Cohors I Delmatarum, based at Maryport (Alauna) in northern England is hardly likely to have solely relied upon Dalmatian recruits, drawn from their homeland 1,000 miles and several provinces away, and so their recruits would have to come from a more local source. That being said, there is evidence that both Thracian and Syrian units continued to bring in men from their homeland, at least for a time, but they may only be a part of the levy.

There is a strong suggestion that there were periodic recruitment drives, some of which may be responsible for inscriptions in Britain mentioning German recruits in units with other tribal names (Vexillatio Germanorum Voredis as part of Cohors II Gallorum, for example). In addition to Petillius Cerialis bringing German units to Britain after Civilis revolt, Tacitus tells us that in the wake of the Boudiccan revolt a decade earlier 'Its strength was increased by the Caesar, who sent over from Germany two thousand legionaries, eight cohorts of auxiliaries, and a thousand cavalry.' It may be that at such times units were merged and used to strengthen depleted forces. There is also the possibility of Mauretania as a source of recruits in the early second century, suggested by the diploma of AD 109 (*CIL* XVI, 161), which seems to indicate units present briefly between other postings, as will be noted in our unit listings later.

Despite the trend, since Vespasian, of posting men away from their homeland, recruitment into the *auxilia* did not rely solely on foreign recruiting drives, and involved local levies increasingly as time passed. Indeed, during the Antonine period, one Nectovelius, son of Vindex, was buried at Mumrills on the Antonine Wall, his gravestone stating that he had served nine years with Cohors II Thracum, and that he had belonged to the tribe of the Brigantes. This is clear evidence that as early as Antoninus Pius, and possibly even Hadrian, recruits were being drawn from the local population. Indeed, this local recruitment is further supported in literature. A second Briton serving in a unit on the island is identified on diploma *CIL* XVI, 130, one Saturninus from Colchester serving in Cohors I Fida Vardullorum. We already noted in the *auxilia*'s history how Agricola had among his army at Mons Graupius in AD 83 a number of Britons, and the *Notitia Dignitatum* notes at Newcastle-upon-Tyne in the closing days of Roman Britain the presence of a unit of Cornovii, who would have been drawn from the area of Shropshire and Cheshire, unless this is a transcription error, which we will investigate later. Clearly, while local recruitment may not be favourable, remembering Civilis and the Batavi, it *was* necessary.

This is not to say that all units followed the same course. As previously noted, Rome was deft at bending its own rules, and evidence from Intercisa suggests that the unit of Syrian archers there continued to draw recruits from their homeland at least from the AD 180s until AD 252, more than two generations of service.

Cheesman delves into this subject, listing recruits stationed in Pannonia. The origins of these recruits are given as follows:

Pre-Vespasianic			**Post-Vespasianic**		
Pannonia	4	Local recruits	Pannonia	9	Local recruits
Thracia	1		Dacia	3	
Dalmatia	1		Thracia	3	
Hispania	1	Into a Spanish ala	Moesia	1	
Narbonensis	1	Into a Spanish ala	Pontus	1	
			Britain	2	Into a British cohort
			Germania	1	Into a German ala
			Palestine	1	

These statistics show a ratio of fifty per cent local to foreign recruitment before Vespasian, against roughly forty per cent afterwards. This is not a huge shift, and it is worth noting that after AD 70, seven of those recruits are from areas bordering Pannonia, and their inclusion as 'local' would bring the proportion of foreign recruits down to twenty-four per cent.

We have already seen terms such as 'cohort' and '*ala*' scattered through the text, which now require a little elucidation. As we have noted, the *auxilia* were often recruited based upon their talents, but auxiliaries of the empire still fit neatly into three categories no matter their skills. The more prestigious of the three unit types was the *ala*, a cavalry unit, the least prestigious was the infantry cohort, and between them lay the *cohors equitata*, a unit of mixed infantry and horse. Archers and spearmen would be found in infantry cohorts, riders either a cavalry *ala* or the horse section of an *equitata* unit. Which type of cohort a given unit was is usually identifiable by its name on inscriptions (examples being *RIB* 1536 with *COH I BATAVOR* as opposed to *RIB* 797: *COH II TRA EQ*). The addition of EQ makes it clear the unit was *equitata*, the absence suggests a straight infantry cohort. The epithets of units will be discussed in more detail later.

Another term we have encountered is 'quingenary', a word that indicates the size of the unit. We know Roman auxiliary units (both *alae* and cohorts) as either 'quingenary' or 'milliary'. Those readers with a grounding in Latin will be able to see in the terms a suggestion that the former denotes a unit of 500, the latter of 1,000. We have to be prepared for a little variation with such terms, bearing in mind that a centurion is known in the Imperial period to have commanded 80 men, despite the word suggesting 100. What we can say with reasonable confidence is that a milliary unit would be roughly twice the size of a quingenary unit. Once again, the size of the unit can often be seen in its name on inscriptions (*RIB* 1580: *COH I TVNGROR MIL* as opposed to *RIB* 430: *COH I SVNIC*). What those numbers actually mean, we shall examine in a little more detail as we now look at the unit types individually.

Model of a Roman auxiliary cavalryman in the Roman Army Museum in Greenhead. (*Author's collection*)

The *ala* takes its name from the Latin for 'wing', which comes from the cavalry's time-honoured tactic throughout the Republic and early Empire, in which they were invariably posted on the flanks of an army (the wings). A rider in an *ala* earned more than other auxiliaries, and their commander, a *praefectus equitum* (or cavalry prefect) outranked his infantry counterparts. Interestingly, while in the infantry differently sized units demanded different ranking commanders (prefects and tribunes), the officer in charge of an *ala* was always a prefect, whether his unit was milliary or quingenary.

We know in Britain of nineteen *alae* either attested or suspected, and of those only one is noted as milliary, but there is considerable difficulty in determining for certain each size of unit's actual numbers. Polybius tells us that in the days of the Republic, a legion's cavalry contingent numbered 300, divided between the command of 10 decurions, giving a number of 30 per *turma* (and a reason for the decurion's title, incidentally).[4] In the *auxilia*, the cavalry *ala* was also divided into *turmae*, each, again, under the command of a decurion. Arrian tells us that a quingenary cavalry *ala* numbered 512 men[5], which would neatly divide into 16 *turmae* (a number confirmed by Hyginus), each with a decurion commanding 32

4. Polybius, *Histories* 25.1.
5. Arrian, *Tactica* 18.3.

men. Certainly the presence of 512 men would suit the designation 'quingenary' quite well, and if we assume the *turma* of 32 includes the decurion and standard bearer, also fits seamlessly with Polybius.

While we may happily accept 512 as the size of a quingenary *ala*, the size of a milliary one causes a few problems. Hyginus states[6] that a milliary *ala* consists of twenty-four *turmae*. If a milliary *ala* follows the same sub-unit size as a quingenary one, then 24 *turmae* of 32 men would make the unit only 768 men. The Coptos tablets (*CIL* III, 6627) give a breakdown of officers and men of 3 *alae*, commonly used to suggest unit numbers, and though they represent only vexillations assigned to a work party, they can still be used to suggest manpower figures. If their numbers are examined, and we are to believe that the *duplicarii* and *sesquiplicarii* (riders on double pay and pay-and-a-half) were commanding *turmae* alongside the decurions, then we find an intriguing possibility. The tablets tell us the 3 *alae* involved in the building work supplied 6 decurions, 1 *duplicarius* and 4 *sesquiplicarii*, giving a total of 11 officers, and 424 troopers. If the troopers are divided more or less equally between those 11 officers, then each turma would contain just short of 40 men. This gives a total, over 24 *turmae*, of 960 troopers. If the officers (decurions and standard bearers) are added to each turma, that brings us to a total of 1,008, a number that fits snugly with the term 'milliary'.

Thus we can tentatively suggest that a quingenary *ala* consisted of 512 men in 16 *turmae* each of 30 men plus 2 officers, while a milliary *ala* consisted of 1,008 men in 24 *turmae*, each of 40 men plus two officers.

To these totals we must add the top brass and the *principales*, the senior men of the unit. Each *ala* was commanded, as stated earlier, by a cavalry prefect. The prefect had a guard, or escort, which was likely drawn from the troopers under his command, and which was commanded by an '*optio*' who ranked as one of the *principales*. Along with him, in that category, would come the *vexillarius*, who carried the *ala*'s colours, and an *imaginifer* (bearing the emperor's portrait on a standard). The *duplicarii* and *sesquiplicarii* would also class as *principales*.

Below the *principales*, among the *turmae*, came the *immunes*, the men whose positions saw them exempt from fatigues, and it is in this category that we begin to find the variation that makes units fascinating. From surviving inscriptions we know that among the staff of the *ala*, each *turma* would have its *signifer* (standard bearer) as mentioned above, but could also have:

- *actarius* – clerk, second to the cornicularius
- *buccinator* – musician with a buccina*
- *cornicularius* – chief clerk

6. Hyginus, *Liber de Munitionibus Castrum* 16.

- *curator* – responsible for provisions
- *custos armorum* – responsible for armour and weapons
- *librarius* – clerk or book keeper
- *stator* – courier or messenger
- *strator* – groom
- *tubicen* – trumpeter*

* Perhaps these terms are interchangeable in an *ala*. Certainly the logistics of attempting to carry and play a bulky *buccina* on horseback suggest as much. The small, portable tuba is much more reasonable for cavalry.

We cannot be sure of the numbers of such positions within each *ala*, and some were clearly only held by one man. Added to this would be the '*beneficarii*' (soldiers seconded for some unstated specific duty that took them out of the normal ranks). We have references to an '*adiutor*' who may have been an adjutant to the commander, to a '*dispensator*', the unit's paymaster, and in one extraordinary

Detail of Trajan's Column, showing horses being transported by boat, the job of a *magister barcariorum*? (*Author's collection*)

inscription from Mauretania (*CIL* VIII, 21568) we learn that Salustius Martialis served in an *ala* as the '*magister barcariorum*', the master of the boats!

Below the commander, the *principales* and the *immunes*, then, the rest of the *ala*'s manpower consisted of the *equites*, or 'troopers'. Given the above figures, I am sceptical about Hyginus's statement that 'there are decurions, double-pay men and pay-and-a-half men in the same number as there are troops.'[7] If this was truly the case, then when the *immunes* are added to the *principales*, there would be precious few men in the unit left to do the bulk of the work.

The infantry cohort will be considerably more familiar to those conversant with the Roman legions, since the model for the auxiliary cohort was drawn directly from their legionary counterparts. Just as in a legion, the quingenary auxiliary cohort consisted of 6 centuries, each of 80 men, giving the cohort a strength on paper of 480 men. When the officers are added to this total, it reaches 498, perfectly in line with the term 'quingenary'. The milliary cohort, Josephus tells us in his *Jewish War*, was 1,000 strong, which fits with the name, clearly: 'Now these ten cohorts had a thousand footmen, but the other thirteen cohorts had no more than six hundred footmen a piece, with a hundred and twenty horsemen.'

Josephus's figures are treated with some scepticism, as they do not tally with those of Tacitus and Suetonius. In fact, archaeological evidence refutes Josephus' statement and, in practice, the milliary infantry cohort seems to have numbered 800, based on the excavations at Fendoch, which revealed 10 barrack blocks, each designed for 80 men. This figure is further borne out by Vindolanda tablet 154, containing a strength report for the milliary Cohors I Tungrorum, which states a full unit strength of 752, though the tablet lists only 6 centurions, which is a puzzle.

The infantry cohort was commanded by either a prefect (in the case of quingenary cohorts) or a tribune (for milliary ones). Of course, as always with the Roman military, while this is a set rule, it was able to be bent or broken as the situation demanded, but for the most part, these two ranks provided the command for the two different-sized units. The cohort would seem to have an *imaginifer*, but no principle *vexillarius*, the standards and banners being carried at century level.

Each century was commanded by a centurion, his second-in-command, the *optio*, and the third-in-command and keeper of the watchword, the *tesserarius*, ranking among the *principales*, again much as their legionary counterparts. Each century would also have a *signifer*, carrying the standard, who ranked among the *principales*.

7. Hyginus, *Liber de Munitionibus Castrum* 16.

Below the *principales*, as with the cavalry, came a variety of positions that held the status of immunes.

- *actarius* – clerk, second to the cornicularius
- *buccinator* – musician with a buccina
- *capsarius* – combat medic
- *cornicularius* – chief clerk
- *curator* – responsible for provisions
- *custos armorum* – responsible for armour and weapons
- *librarius* – clerk or book keeper
- *medicus* – doctor or surgeon
- *mensor* – 'measurer', perhaps a surveyor, although in Justinian's *Digest*, Paternus noted in his *Military Affairs*: 'measurers of grain and their assistants', and given the presence of a grain *modius* at Carvoran, this cannot be ruled out.[8]
- *singularis* – bodyguard, commonly seen with *consularis*, suggesting secondment
- *stator* – courier or messenger
- *tubicen* – trumpeter
- *vexillarius* – bearer of the flag, or '*vexillum*'

As with the cavalry, added to this would be the *beneficarii*, and *adiutor*. With the cohort we also have various other references on inscriptions that sit outside the standard rank system:

- *actor* – mentioned on *RIB* 908, this might suggest a man in temporary command
- *aeditor* – noted at Dura Europos, he may have been in charge of the chapel of the standards
- *circitor* – also at Dura Europos, perhaps indicating a patrol leader
- *explorator* – a scout
- *protector* – perhaps a guard or watchman?

And once again, below the commanders, the *principales* and the *immunes*, came the *milites*, or soldiers.

The organization of the *cohors equitata* is a complex subject. Hyginus gives us direct figures, with the milliary cohort comprising 760 infantry (organized in 10 centuries of 76) and 240 cavalry in 8 *turmae* of 30. The decurion and

8. Justinian, *Digest* 50.6.

standard bearer would bring the *turma* total to 32, of course, and centurion, *optio*, *tesserarius* and standard bearer the century to 80. Hyginus goes on to say that the quingenary cohort is comprised of 380 infantry and 120 cavalry (precisely half the larger version).[9] The problem that arises is that he tells us the quingenary infantry is organized in 6 centuries, which would give us centuries of only 63 men, while he tells us the rest is 'in proportion', presumably meaning the 120 cavalry are in 4 *turmae* of 30. Of course, if there has been a transcription error somewhere over the centuries, and the original text held a V instead of VI, that would again give us centuries of 76 men, which would make more sense. In support of Hyginus's figures, the fort of Birrens, which played host to a milliary *cohors equitata*, could certainly house such a unit.

If we were to assume that the infantry section of a milliary *equitata* cohort were identical to the straight infantry cohort, that would give us a figure of 800 men in 10 centuries. If we further assume a cavalry contingent of 8 *turmae*, each of 32 men, that gives us a total of 256, giving us a full unit strength of 1,056. A quingenary unit of 6 centuries and 4 *turmae* would then total 608, a number eerily close to Josephus's figure ('the other thirteen cohorts had no more than six hundred footmen a piece, with a hundred and twenty horsemen.'[10]) The fact that the barrack systems in a number of the auxiliary forts in Britain do not fit any of these figures throws further confusion on the subject.

In terms of rank and position, we can expect to find the same ranks as in a cavalry *ala* and an infantry cohort as appropriate for each part of the unit, with a *vexillarius* for each of the cavalry *turmae*.

While there may appear to be no difference in this type of unit to the centuries of men in an infantry cohort and the *turmae* of riders in a cavalry *ala*, there are, on closer examination. The riders of an *ala*, as previously stated, were better paid than the riders of an *equitata* cohort, and were clearly of higher status. This may have been reflected in the manner of their duties, for cavalry *alae* seem to have been kept largely for direct military action, forming a mounted combat force, while the riders of an *equitata* cohort were perhaps more often used as scouts, skirmishers, escorts and the like. The lower status of the *equitata* riders to those of the *ala* is made clear in the emperor's own expectations, recorded on the excellent Lambaesis inscription (*ILS* 2487): 'It is hard for the horsemen of the cohorts to make a good impression even by themselves; still harder not to displease after the manoeuvre of the *alae*.' That being said, Tacitus notes that at Mons Graupius in AD 83 Agricola fielded four Batavian units and, apart from

9. Hyginus, *Liber de Munitionibus Castrum* 26–27.
10. Josephus, *Jewish War* 3.4.2.

the exchange of missiles, they were the first units to engage in the battle.[11] The Batavian units known in Britain were *equitata* units, and on that occasion, at least, they were sent proudly into battle.

The foot soldiers of the *equitata* cohort could perhaps consider themselves in a better position than those of a standard infantry cohort, too. After ten years of service in the cohort, they could transfer to the mounted detachments, which would see an increase in pay and status, and certainly better conditions when on the move. There is, then, a sort of cascade effect of status and pay among the *auxilia*, with the *ala* at the top, the riders of an *equitata* unit second, then their infantry compatriots third, with the *milites* of an infantry cohort at the lower end.

Of the ranks and positions for the various units named above, special attention is due to the medical personnel. The subject of medics in the Roman army is one that has been the cause of debate among historians for more than a century, and continues to raise questions. That there were medics in the Roman army is undeniable, and not just in the legions. Clearly there were medics at work in the *auxilia*. No matter what arguments are put forward, the remains of buildings that have been convincingly identified as *valetudinarii* (hospitals) in auxiliary forts confirm fairly unassailably that this was the case, let alone tombstone inscriptions. These buildings will be discussed later, but in short, the excavated examples fit the requirements of such structures, medical equipment has been excavated there, they had running water, individual rooms, and a garden where traces of medicinal plants have been identified. The evidence is sufficiently clear.

The question of how medical personnel were organized remains uncertain. There seem to be two different types of medical personnel recorded among the *auxilia*. The first is a soldier with medical duties. These men (an example being *RIB* 1618 from Housesteads) are often recorded as a *medicus ordinarius*. This was long presumed to be a reference to a medical orderly, though it seems on balance much more likely to refer to a medic within the ranks of the unit. Anicius Ingenuus, on the above-mentioned tombstone, is recorded as *medicus ordinarius* of Cohors I Tungrorum. Such men are perhaps to be equated with the *capsarius*, in essence a combat medic, a soldier with at least rudimentary medical training.

A second type of medical personnel is suggested on *CIL* XI, 3007 from Viterbo. This inscription gives the career of one Marcus Ulpius Sporus, and tells us that he had been *medicus* for Ala [Gallorum] Indiana, who were based in Britain, and Ala III Asturum in Mauretania, but also a private salaried physician in the city of Ferentinum, which has been taken as a suggestion that Sporus was

11. Tacitus, *Agricola* 36.

Detail of Trajan's Column, showing *capsarii* at work. (*Author's collection*)

a citizen with medical skills who had been hired or seconded by the military, rather than a soldier who was trained following enlistment.

Other evidence of medical personnel abounds in Britain, from the altar raised by a *medicus* of Ala I Vettonum at Binchester to the health of his unit (*RIB* 1028), to the Vindolanda tablet 156 that records a *medicus* named Marcus building 'the residence' (a hospital?) Justinian's *Digest* even refers to an '*optio valetudinarii*', clearly a man with responsibility for a fort's hospital.[12]

On the subject of pay, we can say little with certainty. There are a number of sources that give us clues, but nothing that we can state with confidence reveals an exact figure. A lot of the estimates of auxiliary pay are by necessity based upon the pay of the legions, for they are more clearly dealt with in ancient sources.

The starting point for any investigation of pay is Suetonius, who tells us that Domitian 'increased the pay of the soldiers one fourth, by the addition of three gold pieces each year.' Speidel, supported by the figures given by Cassius Dio, considers this a fourth of the *new* figure, which tells us that an increase of three *aurei* was a quarter of the yearly pay, which had therefore, previously, been nine

12. Justinian, *Digest* 50.6.7.

aurei and was now twelve. (Cassius Dio notes 'four hundred sesterces should be given to each man in place of the three hundred that he had been receiving.') We also know that legionaries were paid in three instalments over the year, so that was three *aurei* (seventy-five *denarii*, the denomination we shall use from now on) every four months. Domitian may have changed the system so that there were four payments of 75 *denarii*, but we know that it was later back to three payments, each now of 100 *denarii*.

So before Domitian, legionaries were paid 225 *denarii* a year in three payments of 75, then from Domitian to Septimius Severus: 300 *denarii* in three payments of 100. This is an excellent source, for it gives us definitive figures to work from, for the legions, at least.

It has long been assumed that auxiliaries were comparatively poorly paid, some historians claiming the figure was as low as one third of a legionary's pay. More recently, Speidel has thrown a great deal of light on the subject, basing his calculations on two papyri that seem to list an auxiliary's wage (P.Gen.Lat.1 and P.Gen.Lat.4). The figures he arrives at for the *auxilia* amount to five sixths of the value of a legionary's pay, a far closer margin than previously suggested. This would mean that, before Domitian, an auxiliary would receive 187.5 *denarii* a year in three payments of 62.5, and after Domitian 250 *denarii* in three payments of 83.3. The numbers are faintly headache-inducing, I grant you, but his logic is quite convincing, and if these figures were paid in four instalments, they would number four payments of 62.5, which is far easier. This figure seems to be supported by Masada papyrus #22, in which a soldier claims to have received a stipend of 62 *denarii*. Another soldier on the same papyrus claims to have received only 50 *denarii*, though this may be after some official deduction.

Such a narrow pay gap between the legions and the *auxilia* is intriguing, and goes against conventional wisdom, but it does help to explain a few things. Firstly, after Caracalla's edict grants citizenship across the empire, and the lure of the grant for the *auxilia* no longer has such value, it explains why people were still content to enlist. Secondly, it perhaps explains why there is clear evidence of the *auxilia* being joined by men who were already citizens. Thirdly, it might explain the transfer of soldiers from Legion II Traiana Fortis to an auxiliary cohort in Egypt circa AD 215. And lastly, it explains the career path of the famous Tiberius Claudius Maximus. This man, who served in Trajan's Dacian Wars and is remembered for capturing King Decebalus, gives his service record on his tombstone. He had been a regular cavalryman in Legio VII Claudia Pia Fidelis, rising to become the legate's bodyguard and the legion's cavalry *vexillarius*, but who was 'promoted' to become a *duplicarius* and then a decurion in Ala II Pannoniorum. Unless the pay was a lot closer than previously assumed, this career path would make little sense. Parchment P. Berol 6866 suggests strongly that the same pay scales were still in force in AD 192.

If the recipients of the pay on the two parchments named above were standard ranking auxiliaries, then the five sixths pay scale would result in a *duplicarius* receiving 500 *denarii*, and a *sesquiplicarius* 375.

The next change in pay after Domitian comes with Septimius Severus and can be dated rather specifically to AD 198/199. Unfortunately, once again, we have no record of the actual figures, for the *Historia Augusta* says only 'having first raised his soldiers' pay.'[13] Only a few years later, the army was once again enriched by his son, Caracalla, who Herodian tells us 'promised each soldier 2,500 *denarii* and increased their pay by one-half.'[14] Clearly, Caracalla lived by the advice of his father, whose last words were: 'Be harmonious, enrich the soldiers, and scorn all other men.'[15]

Even if we assume only a small raise under Severus, which seems unlikely given his focus on the army, a raise of fifty per cent under Caracalla is a huge increase. Attempts to suggest a modern equivalent of any of these figures are fraught with difficulties and involve some guesswork, and I shall make no attempt to do so here. I can do nothing better than quote Speidel's conclusion: 'Apparently soldiers in the *auxilia* generally did quite well: many had large sums on deposit, and some of them could afford not only a wife but a concubine as well.'[16] To give perhaps some idea of value, however, David Hendin's *Guide to Biblical Coins* tells us that sometime in the first to second centuries an amphora of olive oil sold for 1 *denarius*, and an ox for 100 *denarii*, while 100 bottles of wine sold for 250 *denarii*. Thus, a Domitianic auxiliary infantryman could spend his year's wage on 100 bottles of wine, although I doubt his commander would recommend it!

A small calculation of even the most conservative figures, then, for a cohort of 480 men on 62.5 *denarii* (which ignores a higher rate for officers and *principales*) tells us that just one quingenary infantry cohort could expect a pay wagon every four months carrying in excess of 30,000 *denarii*, and upward of 60,000 after Caracalla. It starts to become apparent why auxiliary forts had underground strong rooms beneath their headquarters buildings, Vindolanda even going so far as to double the wall thickness of the room above it.

Of course, a soldier's pay went beyond simple disposable income. A man's equipment was not supplied free of charge, and the costs were deducted from the soldier's pay, as is stated plainly by Polybius: 'the quaestor deducts from

13. *Historia Augusta*, *Severus* 17.1.
14. Herodian, *History* 4.4.7.
15. Cassius Dio, *History* 77.15.
16. Speidel, *The Pay of the Auxilia* p.6.

The subterranean strongroom at Chesters. (*Author's collection*)

their pay the price fixed for their corn and clothes and any additional arms they require.'[17] Though Polybius was writing about the early Roman army, this system was clearly still functioning in the days of the Imperial *auxilia*. The same Masada papyrus we saw earlier (#22) tells us rather depressingly 'I received my stipendium of 62 *denarii*, out of which I have paid barley money 16 *denarii*; food expenses 20 *denarii*.' More than half the soldier's paycheque on food for him and his animal alone!

Clearly in some cases a soldier's pay did not meet their needs, for we have ample evidence of men supplementing their wages in various ways, sometimes legally, sometimes less so. In the Arsinoites Papyrus BGU.2.462, a soldier named Gaius Iulius Apollinarius seeks the help of the local epistrategist in collecting rent from a man who refuses to pay up, and Tacitus makes it clear that soldiers were not unknown to have a side income: 'the soldiers were to retain their [tax] immunities except in the case of goods which they offered for sale.'[18] Perhaps the most damning snippet of all comes from the Bible: 'Then some soldiers asked him, "And what should we do?" He replied, "Don't extort money and don't accuse people

17. Polybius, *Histories* 6.39.
18. Tacitus, *Annals* 13.51.

falsely – be content with your pay.'"[19] A further hint at potentially illicit incomes among the *auxilia* comes in the fabulous form of the Carvoran *modius*. A *modius* is a grain measure, and the one found at Carvoran is Domitianic, the unit based there, possibly Cohors I Batavorum, probably using it to collect the tax in grain from local farmers. The *modius* is inscribed, stating that it holds 17.5 *sextarii*, while in fact it actually holds 20.8 *sextarii*, meaning that the poor farmer was being cheated out of more than 10 per cent of his grain. We cannot be sure whether this was a mistake, an official decision, or simply the greed of the *mensor*, but it is tempting to imagine the soldiers side-lining 3.3 *sextarii* of grain from every collection for private sale later.

The Carvoran *modius*, in Chesters Museum. (*Author's collection*)

We have already seen the way Rome adopted non-citizens into the *auxilia* in a manner that made use of their unique talents, but it is perhaps worth exploring some of these specializations in the sphere of Roman Britain.

One of the most unique of British auxiliary units is Cohors I Hamiorum Sagittariorum. For full details of this cohort please refer to the unit listings later in this work, but suffice it to say that this is one of the most interesting in the province. Hamiorum tells us the unit was raised in Hama (Syria), while Sagittariorum tells us they were archers. Syrians were noted in the Roman era for their archery, one of a number of Middle Eastern peoples so famed (one might remember the infamy of the Parthian horse archers from whom we derive the phrase 'parting shot') and the Syrians supplied a number of units to the *auxilia*. This is the only unit of archers known in Roman Britain, though, and what can be reconstructed of their activity suggests that their skills were so prized that they were used to supplement other units.

This makes a great deal of sense, for garrisoning a unit of archers together in one fort fails to make use of such a skill, and also leaves other forts without the benefit of their talents. Archers are at their most useful in support of infantry. Of course, undoubtedly infantry and cavalry units would boast a few men who could use a bow, and its training seems to have been a standard feature, according

19. Luke 4:14.

to Vegetius, but our evidence does tend to point to I Hamiorum being spread around the region, vexillations attached to other units. Indeed, the more we learn about the way the Roman army worked, the more it becomes apparent that it was common for units to be split up and distributed across more than one site. Evidence for such appears time and again in the listings later in this book.

Archers are not the only interesting specialization, though. The use of sailors in Roman Britain is a subject worthy of its own book, given the history of the Classis Britannica that supplied us with our own breakaway emperor, Carausius, colloquially remembered as a 'pirate king'. In my work on the career of the great general Gnaeus Julius Agricola, I noted more than once his reliance upon the navy, something unusual in first-century Roman commanders. Given Britain's island nature it is perhaps understandable that the sea and the rivers are of such import, and in addition to the navy, Britain seems to have housed a number of land-based military units of a nautical nature.

Rome has something of a history of forming land units from the navy. In Suetonius's *Life of Galba*, we are told of 'some marines whom Nero had made regular soldiers'.[20] A similar origin can likely be claimed for both Cohors I Classica and I Nauticarum, and Cohors I Menapiorum Nautarum were clearly recruited from among the sailors of the coastal Menapii for their skills, given that those skill are proudly proclaimed in the unit's name. Even towards the close of Roman Britain we find waterborne units attested, and while we can say nothing about the origin of the Numerus Barcariorum, the similarly-titled Numerus Barcariorum Tigrisensium were clearly boatmen brought from the River Tigris, nominally outside the empire, to serve in Northern Britain. That the skills of such units were being utilized is evident not only in their name, but also their deployment. With I Classica seemingly based somewhere on the Cumbrian coast, Barcariorum likely at Lancaster on the River Lune, and the Tigris boatmen on the Tyne at South Shields where more than half the length of Hadrian's Wall was reachable from their supply base by boat, their value is clear. And finally, on the subject of water, we have already examined at length the skill of the Batavi at crossing rivers with their horses while armed and armoured, but it is appropriate at this juncture to recall their particular talents.

Scouts are an interesting subject. It seems likely that every unit had its scouts (we find references to '*exploratores*' within units), and there are ample references to the Romans relying upon native scouts, but did they have dedicated trained scouts of their own? While we do not see entire *alae* or cohorts of scouts in our listings, we find a *numerus exploratorum* at both Risingham and High Rochester in the third century, and by the end of the fourth, also at Bowes and Portchester,

20. Suetonius, *Life of Galba* 12.

suggesting that small units of dedicated scouts were being formed and assigned to specific forts, beyond any in-house scouts.

In general, cavalry were raised from peoples already famed for their riders. Of the nineteen known cavalry *alae* in Britain, nine were drawn from the Gauls and the Germans, whose riders Caesar so praised, and three from the Iberian tribes, whose horsemen were so renowned that one of Rome's famed military manoeuvres, the Cantabrian Circle, was named for them. Going beyond recruiting skilled horsemen into cavalry units, though, Rome's military in Britain reveal more specialized riders still.

We have already briefly met the Sarmatians earlier in this book, and we will be looking at them in a little more depth, but in short the Sarmatians were a confederation of horse tribes that had migrated from the Eurasian steppe to settle in the regions of modern Hungary and Romania. Valerius Flaccus tells us:

> a fierce band of Sarmatians came thronging with savage yells: stiff are their corselets with pliant mail, and such too the armour of their steeds; but stretching out over the horse's head and shoulders the fir-wood shaft, firm resting on their knee, casts a long shadow upon the enemy's field and forces its way with all the might of both warrior and steed, easily gathered up and set in rest again, and once more as swift to go through the midst of the foe.[21]

Ammian goes into considerable detail about the Sarmatians, echoing Valerius Flaccus's description of their equipment, and also lauding their horses and horsemanship. The Sarmatians are part of a long tradition of fierce horse peoples from the steppe, which also included the Scythians and the Huns, and their value is therefore clear. The far-reaching effects of introducing Sarmatian auxiliaries into northern Britain may be alluded to in this extract from the Fell Pony Museum website:

> On retirement the Sarmatians may have received land in Bowland, as a tradition of horse-breeding survived in the area through to the monastic period. Kirkstall Abbey had horse breeding 'ranches' in the Slaidburn area up until the dissolution. A local farmer has deeds for his farm that actually mention that horses were kept rather than sheep or cattle as they were able to escape the predations of wolves. (D Higham, pers. comm. 2002, quoting the opinion of Dr M C Higham).[22]

21. Valerius Flaccus, *Argonautica* 6.231–238.
22. https://www.fellponymuseum.org.uk/fells/rom_dark/cavalry.htm.

Two sections of Trajan's Column, both showing heavily armoured Sarmatians in action. (*Author's collection*)

The Sarmatians, then, were fierce barbarian warriors, fully armoured and fighting with a long lance, the heaviest cavalry Rome fielded for centuries. We will examine these extraordinary and extremely foreign recruits in more detail later. They may be the most exotic horsemen we had in Britain, but beyond these shores, we also find *dromedarii* (camel cavalry) among the *auxilia*!

One of the types of unit known since the early days of Rome and yet unattested in the imperial auxiliary is slingers. We know that the slingers of the Balearic Islands were prized as troops, and were used by Rome in the Jugurthine War, and slingers are portrayed on Trajan's Column, suggesting that they were still employed in war by Rome into the second century. That there are no dedicated sling-based units, of course, does not mean they were not around. From the time of Trajan, as we've already seen, irregular units termed '*numeri*' came to be employed, and so it is possible that sling units were present as *numeri*. However, in Britain at least, it seems likely that slings were simply one of the weapons in which auxiliary units were trained. Vegetius tells us: 'Recruits are to be taught the art of throwing stones both with the hand and sling.'[23] This training is illustrated in Britain by finds of large numbers of sling bullets at auxiliary forts

23. Vegetius, *De Re Militari* 1.16.

such as Caerhun, Strageath, Ardoch, Abergavenny and Ambleside, but most particularly in the 130 examples found at Burnswark Hill fort, a site that appears to have been used as a training camp for sieges.

One other unit type seen in Britain is referenced on no less than five inscriptions from Risingham, Great Chesters and Cappuck. Each of these mentions a vexillation of *gaesati* (spearmen) from Raetia. That they are specifically referenced by the weapon they bore suggests a prized specialism. See the unit listings for details.

Nowhere is the specialization of auxiliary units more evident than in the late Roman work, the *Notitia Dignitatum*. Some would argue that the units in the *Notitia* should not be equated with the *auxilia* of the earlier Empire, but the inclusion of known auxiliary cohorts and *alae* in the listings, and the clear purpose of the units, which mirrors the role of the *auxilia*, is sufficient in my opinion to include them. No work on the *auxilia* of Britain would be complete without reference to the last known source of information on the island's units. Thus, by the end of Roman control in Britain, in addition to unit types described above, we also find references to:

- *Catafractarii* – possibly to be equated with the Sarmatians (more on this connection later), a catafract (also cataphract) is a type of later Roman heavy cavalry in which both horse and rider are fully armoured, forming something akin to the Norman knight.
- *Directorum* – We cannot be certain whether this is a reference to the aiming of a missile weapon, or a duty controlling the dispersal of messages and commands.
- *Vigilum* – Watchmen. Interestingly, given the profusion of signal towers, watch towers, shore forts and the like in late Roman Britain, the only unit of which we know that bears this name was based at Chester-le-Street, a location that does not immediately leap to mind as in need of watchmen.

Among the many generalizations and assumptions about the Roman army that have pervaded the general consciousness for the past few centuries is the belief that the Roman legions built the great infrastructure of the empire. That the legions built forts and fortifications, roads and aqueducts, bridges and monuments is hard to deny, of course, for the evidence is there to be found. A glance at the reliefs on Trajan's Column, for instance, shows legionaries (identified on the column by the wearing of *lorica segmentata*) building forts and bridges. Milestone IAA 1994–1682 in Jerusalem names the erectors as Legio X Fretensis. Cassius Dio tells us of Varus in Germany: 'the Romans, even before the enemy assailed them, were having a hard time of it felling trees, building roads, and bridging places that required it.'[24] Caesar, Trajan and Constantine all built famous bridges, and all would have been constructed by their field army, principally the legions. As far as aqueducts go, one only has to read the wonderful tale of Nonius Datus of Legio III Augusta and his trials and tribulations with a failed aqueduct in Africa to see legionary work illustrated. In Britain, evidence of the legions working on Hadrian's Wall and the Antonine Wall is abundant, and it is a generally held opinion that the legions also built the forts in which the *auxilia* resided, based on the many building inscriptions. Legio II Augusta is recorded with work at places like Benwell and Chester-le-Street, IV Victrix at South Shields and Piercebridge, XX Valeria Victrix at Maryport and Vindolanda, and these are just a handful of sites plucked from the mass.

But Rome could not have relied upon the legions as their only force constructing important works. Before the existence of a standing army, private contractors and slave labour were the prime force in public works in the cities, structures required by an army on campaign falling to them to create. Of course, much of Rome's provincial construction was driven by military need. Roads in particular were important in order to bring forces and resources from the hinterland to the frontier, and to secure an efficient supply system. Of course, not all roads were paid for by the government, and many of the smaller routes will have been private constructions. Still, epigraphic evidence and ancient sources together tell us the legions worked on much of the empire's infrastructure.

Yet the legions constituted less than half the military manpower of the island. Tradition may hold that the legions were the builders for Rome, but plenty of evidence exists to show they were not alone. Manchester has produced a series of centurial stones of Cohors I Bracaraugustanorum, each for a different centurion, heavily outnumbering evidence there for the legions, which strongly suggests that the cohort took the lion's share of the work when the fort was rebuilt in stone. Evidence for auxiliary units building, reconstructing or repairing parts of forts is plentiful, too, and not always their own fort. High Rochester has produced dedications for building work by Cohors I Vardullorum, I Lingonum, IV Gallorum, and II Nerviorum, the last two together on one slab (*RIB* 3491), showing that they shared the work on whatever building they erected. In fact, IV Gallorum and II Nerviorum may have been paired often for projects, since dedications by both are also attested at Risingham, and at Chesterholm. A lovely example of such work comes from Birdoswald (*RIB* 1912). Though it does not tell us which cohort was involved, it tells us what they did: 'the … Cohort restored the commandant's house, which had been covered with earth and had fallen into ruin, and the Headquarters Building and the bathhouse.'

24. Cassius Dio, *History* 56.20.

Building inscription for Cohors I Tungrorum from Castlecary on the Antonine Wall, now in the Hunterian Museum. (*Author's collection*)

Cohors IV Breucorum takes this construction even further, for at Slack and the nearby tile works of Grimescar Wood many tiles stamped by the unit are in evidence, as well as at sites far from their garrison, such as Castleshaw, Castleford, Manchester and Aldborough.

So even if it was the legions that constructed the initial phases of the forts across the province, there is ample evidence that the *auxilia* were involved in stone rebuilds, reconstruction, renovation and care of the forts. There is even possible evidence in the form of *RIB* 2014 that the *auxilia* took part in the construction of Hadrian's Wall, since it would appear that Cohors IV Lingonum were responsible for part of the vallum.

Thus far, this is perhaps no great change in perception. Auxiliary units were assigned as garrisons to these places and so their work on upkeep is to be expected. Yet if we look a little further afield, outside the forts, we find other interesting evidence. Cohors I Cugernorum set up a milestone not far from their base at Cramond, suggesting their involvement with work on the road that ran from there to Newstead.

Aqueducts are an interesting case. There are building inscriptions recording work on aqueducts at Caernarvon, Brougham, South Shields and Chesters. In each case, the auxiliary garrison is referred to in the dative case, indicating that the aqueduct was constructed *for* them. However, in each case the subject responsible for the work is not given:

- Caernarvon (*RIB* 430): 'restored the channels of the aqueducts fallen in through age for Cohors I Sunicorum'

- Brougham (*BRIT.* 40.4): 'water-supply for the Cohors III Bracaraugustanorum'
- Chesters (*RIB* 1463): 'Water brought for the Ala II Asturum'
- South Shields (*RIB* 1060): 'brought in this supply of water for the use of the soldiers of Cohors V Gallorum'

However, *RIB* 1049 from Chester-le-Street has been restored as: 'troopers of the Ala [...] Antoniniana [...] brought in a water-supply, and erected a bath-building from ground-level.' Given the epithet 'Antoniniana' it seems undisputable that the named unit is Ala II Asturum. The inscription is datable to AD 216 and the *ala* was apparently based at Chesters then, which suggests that Ala II Asturum sent a work party from Chesters to construct the aqueduct channels and baths at Chester-le-Street. Since this appears to be evidence of auxiliary cavalry working on a water supply, it may be possible that the other works above mentioned were also built by auxiliaries.

Such evidence of construction by the *auxilia* leads us to draw another interesting conclusion. We have already seen how the *auxilia* could boast medical personnel in parallel to the legions, but if such construction work was carried out by auxiliaries, then that suggests that auxiliary units also contained the necessary skilled engineers to oversee such work. It is possible that the '*mensor*' recorded at Chesterholm was a surveyor rather than a grain measurer (the term could easily apply to either), but irrefutable evidence of this came to light only relatively recently, in the excavations of the bathhouse at Binchester, which unveiled an altar set up by a soldier of the Ala Vettonum who is described as an '*architectus*'.

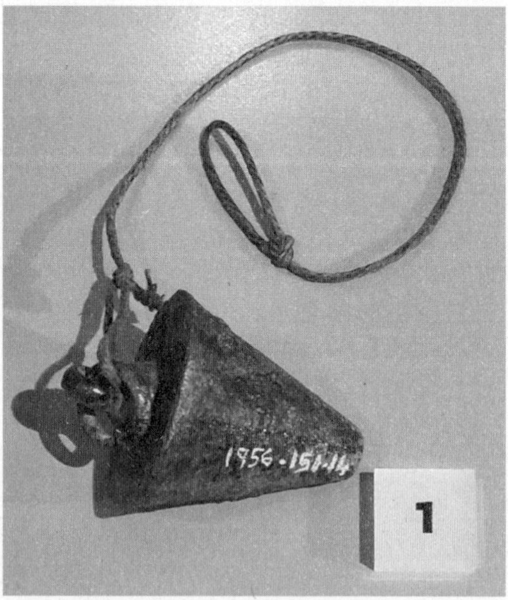

A bronze plumb-bob from Housesteads, perhaps belonging to an engineer or *architectus* of Cohors I Tungrorum, in the Hancock Museum. (*Author's collection*)

The duties of the *auxilia*, therefore, are now seen to be varied. The longstanding notion that the legions were moved around where any skill was required, while the *auxilia* sat in garrison, is plainly erroneous. The *auxilia* were employed in construction, not only of their own garrisons, but were brought in for building projects at the forts of other units. They worked on roads and milestones. They were involved in manufacture and distribution. The tiles from Slack that have been found across the north of England were the work of an auxiliary cohort. It may be that the leatherwork that seems to have abounded at Catterick was a military endeavour as much as a civilian one, for the letter asking for a cart load of hides from there (Vindolanda tablet 343) was a document burned among the military records in the fort, and therefore almost certainly belonged to a soldier. Construction and manufacture clearly occupied the time of auxiliary garrisons.

In addition to this, in other sectors there was an element of control required, which will have been supplied by the *auxilia*. Swaledale, for example, was a concentrated supply region for lead, and the mining operations up and down the valley were of great import to the military as well as to any civilian operations. The mining site at Whitcliffe Scar displays not only adits, but also what appear to be two small fortifications, at least one of which may have housed a small detachment of soldiers. Certainly, the fort at Catterick, at the bottom end of Swaledale where it meets Dere Street, will have acted as a control centre for the shipping of lead in and out of the dale.

Clearly one main purpose of the *auxilia* was to provide a garrison to impose control over an area, a job that becomes more important the closer to the empire's borders those garrisons get. But even in the hinterlands the garrison's duties would include administering the area – policing, as it were. Forts were often built at important road junctions, at the crossing points of waterways, and the like, and Roman control at such sites was vital. Even in late Rome, when the province of Britain far back from the northern border and the shores must have been settled and peaceful, we find a unit of scouts based at Bowes, likely to patrol the Stainmore Pass and prevent brigandage.

And despite the general focus on the legions, and the notion that the *auxilia* existed to *support* the citizen troops in times of war, that is clearly only part of the story. Early on we have ample record of the *auxilia* involved in the initial stages of invasion, in combat in Anglesey, in Scotland under Agricola, and it is worth remembering that at Mons Graupius in AD 83, Agricola committed only *auxilia* in one of the most important fights in Romano-British history, keeping the legions in reserve. That the *auxilia* continued to play a military role during the troubles at the start of Hadrian's reign, in Antoninus's advance to the Forth-Clyde line, in Severus's war in Scotland, in Constantius's invasion of Scotland, and in numerous other smaller or unrecorded incidents, is suggested by the number of honorific and imperial titles given to the units, to their grants of citizenship, their receipt of torques and so on. Indeed, even a cursory glance at the battle scenes on Trajan's Column shows just how involved the *auxilia* were in Rome's wars.

Detail from the Portonaccio Sarcophagus showing a battle scene c. AD 200, including legionaries and *auxilia*, an auxiliary horsemen in scale armour clearly visible in the foreground, in Museo Nazionale Roman. (*Author's collection*)

Perhaps, in light of all of this, it should no longer surprise us that an auxiliary might receive almost as much pay as a legionary, and that citizens might want to enlist in the *auxilia*.

All that remains in this discussion, then, is to detail the various titles and epithets we see applied to auxiliary units, both *alae* and cohorts. In this section I have not included regional names (e.g. Asturum) or names that appear to record an early commander (e.g. Indiana).

Little explanation should now be required for '*Milliaria*' or '*Equitata*' when found in a unit's name, but there are a number of others that appear time and again:

- Civium Romanorum – This title confirms that at one time the unit was singled out for imperial honours, for some exemplary service. On that occasion every member of the unit received Roman citizenship immediately, though this did not apply to men recruited after the grant.

- Fida – Literally 'Faithful'. This epithet seems likely to have been applied to units who remained loyal to an emperor during a crisis. The only instance of its use in Britain is on Cohors I Fida Vardullorum, who may have acquired it for their support of Vespasian in AD 69.
- Herculeana – A reference to Hercules, this epithet is unlikely to directly reference the god, but more likely Emperor Maximian, who took on the title Herculius during his co-rule with Diocletian (Iovius). As such, the unit was probably either raised or enfranchised by Maximian.
- Liberorum – This is an uncertain translation. For a full discussion on the word, see the listing for Cohors II Tungrorum. If read correctly, then it denotes a unit formed of free citizens, something unique among the *auxilia*.
- Miniata – Again only found on one cavalry *ala*. The word suggests 'painted red', which might be a reference to some custom among the unit, either in war or in the *hippika gymnasia*.
- Ob Virtutem Appellata – Literally 'Called the Virtuous'. This only references one unit, and there is a great deal of confusion surrounding its identity and that of another unit (listed separately in this work as Ala Augusta Gallorum Proculeiana and Ala Augusta Ob Virtutem Appellata, though which may in truth be the same unit). As such, we cannot say with any confidence how the unit came to be so 'virtuous', although it seems likely that they remained loyal to an emperor in a time of trouble.
- Pia Fidelis – 'Pius and Faithful'. Much like Fida and Virtutem, this title would be awarded to a unit who remained faithful during dangerous times. Of all the epithets here, *Pia Fidelis* is one of the most common, applied to several units in Britain, as well as units abroad, and also to legions.
- Praetoria – Since this epithet is only found on one unit (Ala I Hispanorum Asturum), there is no parallel to work from. It has been argued, though largely discounted, that the PR stood for Prima, but Praetoria seems, as Davies suggested, the likely candidate. Davies proposed that the Praetoria was an honorific gained during the Severan campaigns in Scotland. If so, it might imply that the unit acted in some sort of guard or escort manner to the emperor. The unit certainly held it during the time of Gordian III, and there is no reference to it prior to the third century. It is interesting to note a line in the *Notitia Dignitatum*: 'In the Britains with the worshipful count of the Britains: One *palatine* auxilium, one legion of the line.' It is tempting to make an educated guess at this. Only two legions are mentioned in Britain in the *Notitia*. One is II Augusta, who seem to now be in a garrison role at Richborough, but VI Victrix seems to still be based at the capital in York, and tops the list of the

Dux Britanniarum. This could be the *Notitia*'s 'legion of the line', and if we were to look at the phrase '*palatine auxilium*', and see the *palatini* as Constantine's replacement for the praetorians, then the only auxiliary unit named Praetoria fits the bill snugly.
- *Sagittariorum* – As noted in the text, this word identifies the unit as archers.
- *Torquata/Bis Torquata* – A torc/torque is a decoration won in battle. It is uncertain how the decoration was granted and kept, but likely the torc was appended to the unit's standards. The receipt of the honour was noteworthy enough for it to then be added to the unit's name. 'Bis Torquata' refers to receiving torcs twice, clearly an extraordinary thing.
- *Veterana* – This is not a reference to veterans, as it might at first appear, but more a reference to age. It is often applied when two units with the same name are serving in the same theatre, the older of the two adding 'Veterana' to distinguish itself.
- *Victrix* – Literally 'Victorious'. Not an uncommon epithet, this one is also seen among legions, such as VI Victrix, based at York. The reason for the acquisition of the name seems straightforward.

Finally, units also acquire the names of emperors, either because they were first raised on that emperor's orders, or because they distinguished themselves under him, or possibly at times simply in order to show outstanding loyalty to him. Among the imperial epithets found in British units we see (in chronological order):

Domitiana
Ulpia/Traiana
Aelia
Antoninianae
Severinae Alexandrinae
Maximiniana
Gordianae
Postumiana
Tetricianorum
Probiana

A number of units take on the word Augusta, referencing the emperor without his specific name. There is a suggestion that at least in some cases these are units that had taken on 'Flavia' in the reign of Domitian, but dropped it like a hot potato on his death and damnation, shifting to a much safer and non-aligned 'Augusta' instead. 'Domitiana' appears only on one altar from Brohl.

Where a unit can be seen with Nervana or Nerviana, this should not be associated with Emperor Nerva, as such epithets usually take an emperor's family name, rather than cognomen, and so under Nerva they would be more likely to be named Cocceia. As such, this is more likely a reference to the Belgic tribe of the Nervii.

Chapter 3

Provinciae

(Origins and ethnic persistence in the *auxilia*)

With any examination of the *auxilia* there is the temptation to scurry down the rabbit hole of ethnicity. The throwing together of men from such disparate lands creates a fascinating subject, for the ethnic variation in the Roman *auxilia* goes far beyond anything even the European Union could hope for today. In fact, the subject is worthy of its own book, and fortunately it's been written. Ian Haynes's excellent *Blood of the Provinces* covers that very subject in huge detail, and so I do not intend to reinvent the wheel. I recommend that tome for anyone who wants to delve deeper than I have the space for here.

That being said, this book would be lacking if I did not cover the subject to some extent. I myself had never really considered the importance of ethnicity in the *auxilia* until recently. At Park in the Past, in Cheshire, re-enactors have begun to reconstruct a Roman fortlet, and on one weekend in 2023 an event was held there to celebrate the Roman auxiliary. The moment I walked in I was struck by the ethnic issue. I attended on only one day of the week-long event, and yet in that one day alone I was fascinated by the diversity. Walking through the gate I was greeted by Tungrian auxiliaries in gear that looked very familiar, which one might consider a 'standard' auxiliary kit. But on the ramparts was a Dacian archer. Standing close to the *fabrica* block was a yurt, outside which a Thracian centurion stood, his kit a fascinating blend of the Roman and the non-Roman. This, then, was the world of the *auxilia*. Riding through the fort itself was a Sarmatian, complete with composite recurve bow. More Dacians were gathered in the bright sunshine at their tent, discussing their armament, which was only familiar to me as I had visited Romania and studied Dacian finds, and moreover they were discussing them in Romanian, for this was a re-enactment group from Romania.

The ethnicity of auxiliary units seems apparent even at just a glance, for the names of cohorts and *alae* often betray their origins. In fact, barring those *alae* whose title seems to have been born of their first commander's name (Indiana from Julius Indus for example), the rest are almost entirely named for tribes, cities or regions. Notably those units raised in the West are often tribally titled, those in the East named for cities or geopolitical areas. This simply reflects the worlds they inhabited, for the West was still largely a tribal culture, while the East had evolved into city-states centuries earlier.

Dacian warrior as portrayed by reenactor of Terra Dacica Aeterna. Note the distinctive helmet, the shield identical to those portrayed on Trajan's Column, and the draco standard of the Dacians and Sarmatians. (*Author's collection*)

Of course, the importance of this over time can be overstated. We have already seen how, as the decades rolled by, units began to recruit from their local regions, as well as receiving manpower from external recruitment drives. As such, there is a very good chance that by the time of the Third Century Crisis, a unit like Cohors II Asturum had very little, if any, actual Asturian blood in its ranks.

But that need not matter. Any look at the modern military shows distinct parallels. The US 1st Cavalry Regiment proudly bear their name, despite the fact that the last man to have ridden a horse in the unit is long gone. In my own quiet corner of Yorkshire, the town of Richmond houses a museum to the Green Howards Regiment, the name still spoken of with pride, despite having been incorporated into the 'Yorkshire Regiment' in 2006. They acquired their name under Charles Howard in the eighteenth century, and any real connection to the Howards has been gone for centuries, but the name is still important. It still carries pride. In many ways it came to define the regiment, to give it an identity. Need we look any further then for an explanation of the importance of an auxiliary unit's name? There may have been no Spaniards in II Asturum by the third century, but an *imaginifer* by the name of Gemellus went to the expense of raising an altar at Brohl on which his unit is named COH II ASTURUM P F D (Pia Fidelis Domitiana). Indeed, the increasing addition of epithets to a unit's name only further illustrates that pride which drives a man to spend good money having extra words carved on altars. I suspect the stone mason at Benwell was quite tired by the time he'd finished ALA I HISPANORUM ASTURUM PUPIENAE BALBINAE GORDIANA. At Housesteads, as he forked out coin after coin to the mason, the German officer Hnaudifridus must have been cursing that his parents didn't call him Bob.

I digress, but the point is made. In the same way as soldiers from the very fringe of empire, whose own ties to Rome were largely financial, proudly carried the image of the emperor and bore his name, their own unit's name became an institution in itself, above any geopolitical meaning.

So who were these people who formed the *auxilia*?

One thing that needs to be borne in mind throughout is that any investigation of ethnicity is largely reliant on what can be unearthed from the ground or lifted from ancient texts. We are often limited to imagery, archaeology and names, and absence of evidence is not often evidence of absence. We can only report on what has been found. As such, some peoples' cultures were close enough in imagery to those of Rome and Britain that any evidence of their ethnicity is swept away. We find it hard to see clear evidence of Gallic or Germanic culture, for example, for their armament and mode of dress had more or less aligned with most of Rome's military, meaning that they entirely fail to stand out. Only those who clung to the unusual can be easily spotted.

One of the more obvious of these is the Moors. The Mauri, as later Romans called them, their land being Mauretania, were a people whose origins lay in the Numidians of Republican days, in the Berber peoples and the semi-Nomadic raiders of the Saharan fringe. Their territory stretched from the Atlantic through what is now Morocco (note the origin of that name in passing) and Algeria, to the edges of Tunisia – quite a wide swathe of land, some 800 to 900 miles across.

They were a hard people, as is made clear by their history and by the few records we have of them. Strabo tells us:

> the nomads of today not only excel in the skill of hunting (and the Romans take a hand in this with them because of their fondness for fights with wild animals), but they have mastered farming as well as the chase.[1]

And of their fighting style, Herodian gives us 'the Moorish hurl their javelins from a distance and attack and retreat nimbly.'[2] The Mauri were light cavalry, unarmoured and bare-headed, utilizing light javelins, a combination that made them fast and agile, able to cause havoc and be out of danger in the blink of an eye – a skill that was of great value to Rome. And Rome had learned this at their cost, for the Mauri were trouble. From their inclusion as a province in AD 44 until the turn of the seventh century, long after the fall of Rome, every

1. Strabo, *Geography* 2.5.33.
2. Herodian, *Roman History* 6.7.8.

Distinctive Mauri riders on Trajan's Column. (*Author's collection*)

century saw incursions into Iberia by the Mauri, raids brutal enough to have been recorded by Roman historians and to have drawn the military attention of successive emperors.

So what does one do with such a dangerous and restive force? The answer might simply be 'hire them, take them far from home, and point them at someone else who's causing you trouble.' Certainly, despite the repeated difficulties they caused for Rome in the West, they were prized in the army. Trajan valued their service in Dacia enough that they are commemorated on his column.

Emperor Macrinus was proud of his Mauri heritage, as Cassius Dio tells us:

> Macrinus was a Moor by birth, from Caesarea, and the son of most obscure parents, so that he was very appropriately likened to the ass that was led up to the palace by the spirit; in particular, one of his ears had been bored in accordance with the custom followed by most of the Moors.[3]

By the time of the *Notitia Dignitatum*, dozens of Mauri units fill its pages. Their presence in Britain is attested as a unit at Burgh-by-Sands, as well as the Mauri slave of an Iberian soldier based at South Shields.

We have already mentioned Hamian archers, brought from the city of Hama in Syria, and we know of only that one Syrian unit based in Britain, but the value of these men is clear. Once again, they seem to have played an important part in the Dacian Wars. There have been suggestions that the archers shown on Trajan's Column may not be Syrian, but evidence weighs in their favour, as I will explain later when we look at auxiliary equipment. The Syrians supplied cavalry and infantry to Rome's *auxilia*, but they are overwhelmingly known for their archers. Of the twenty-two units recorded as originating in Syria, eighteen are archers. Of course, the East was famed for its bowmen more than anywhere else in the empire, barring perhaps Crete. The Assyrians had fielded celebrated bowmen, and the Parthian horse archers secured a place in history with the phrase 'parting shot' derived from their skills. Syrian archers are generally identified as wearing conical helmets and carrying a distinctive shape of recurve composite bow. Their equipment will be discussed later.

It might be noted how Vegetius tells us that the army trained its men in the use of both bows and slings, and one might then wonder about the value of a unit of archers drawn halfway across the world, and about the sense of forming them into a specific unit. The answer is probably a simple question of skill. A football team does not overlook the option to buy in a world-class player just because they already have eleven men, each with two legs and who knows how to kick a ball.

Rome's links with Syria grew ever closer through the late-second and third centuries. Septimius Severus's wife was from the Syrian city of Emesa, and from then on until the fall of Alexander Severus, Emesa and Syria were at the very heart of Roman culture. Thereafter, the city-state of Palmyra became a focus for Rome, sometimes as an ally, sometimes an enemy. The adoption of Sol Invictus into Rome's pantheon has its roots in Syria, and Syrian military evidence remains in Britain in the form of the Hamian archers at Carvoran, at Kirkby Thore, and perhaps at Housesteads, but beyond that, Syria is also evidenced in one of Roman Britain's most fascinating tales.

3. Cassius Dio, *Roman History* 79.11.

A man by the name of Barathes was buried at Corbridge. His tombstone (*RIB* 1171) names him a native of Palmyra and tells us that he was a *vexillarius*, the flag bearer of a military unit. It is distinctly possible that he had been *vexillarius* for Cohors I Hamiorum, who were based 20 miles away at Carvoran. But Barathes is, literally, only half the story. While in Britain, a man called Barates from Palmyra acquired a slave woman by the name of Regina from among the Catuvellauni in the region of Hertfordshire. In time he freed and married her. Barates buried his wife at South Shields, again around 20 miles from Corbridge, and he is mentioned on her gravestone (*RIB* 1065). It is possible that these two men are not the same, but the temptation to put them together into a full story is strong: Barates the Syrian flag bearer and his ex-slave British wife. Together or separate, though, they illustrate the extent to which Syrians had integrated in the north of Britain.

Indeed, the eastern influence does not end with Syria. An altar at Corbridge raised to the god Hercules of Tyre shows another Levantine connection, though it is not impossible that the same Barathes commissioned the work. The altar (*RIB* 1129) is in Greek (the lingua franca of the East) and was set up by a priestess by the name of Diodora (again, a Greek name, meaning 'Beloved of God', the god in question being Zeus).

Regina and Barates were not South Shields' only links with the East, and we've already mentioned the late garrison of that fort: Numerus Barcariorum Tigrisensium – boatmen from the River Tigris in modern Iraq. Their integration

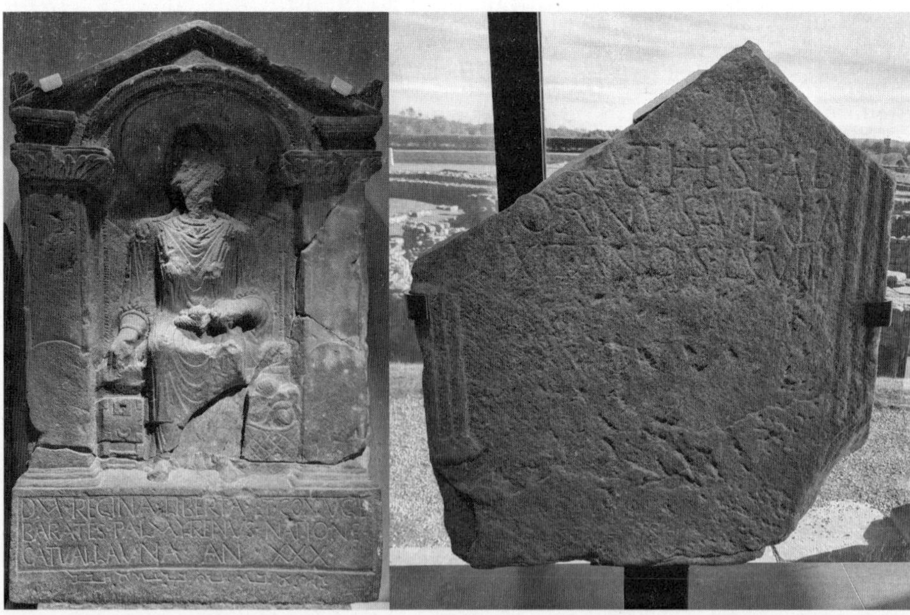

The tombstones of Regina at South Shields and Barathes at Corbridge. (*Author's collection*)

into the region is made clear by the fact that South Shields fort, which seems to have been called Lugudunum at the turn of the third century, is known in the *Notitia Dignitatum* as Arbeia – the Place of the Arabs. In fact, these various eastern examples in the northeast of England elegantly illustrate something we will revisit later: the fact that in bringing these cultures into the empire via the *auxilia*, Rome in fact opened itself up to their influence, and just as Rome enveloped those various cultures, they also inevitably put their stamp on Rome in return.

Rome's border for most of its history stopped in the east around the Euphrates River, the Tigris only just touching on the empire, and so the 'Arabs' in question were almost certainly not raised within Roman lands. If it seems strange that the Roman military might employ a unit drawn from outside its borders, it is worth noting that the Tigris boatmen are far from alone in this. Indeed, it is quite possible that the gradual inclusion, enfranchisement and Romanization of the bulk of the *auxilia* is what set emperors from the second century creating irregular *numeri* and units from ever-more-barbarous and foreign sources. Though they were never stationed in Britain, other Parthians from outside the empire served in the *auxilia* in the form of the Ala Parthorum and the Cohors Parthorum, units that seem to date from the era of Augustus. And the Parthians were not unique, either. The Mauri, after all, were nominally within the empire's borders, yet they recognized no such boundary, spreading out into the semi-arid fringe of the Sahara. And, of course, while we are looking at ethnicities far removed from Romano-Britons, there remains one fascinating group drawn from outside the empire to serve in the *auxilia*, and who we have already mentioned several times: the Sarmatians.

Of all the groups supplying men to the *auxilia*, the Sarmatians may be the most interesting. We are only aware of their recruitment on that one occasion, following the wars of Marcus Aurelius in AD 175, but the fact that 5,500 of them were sent to Britain is enough to make them one of the greatest suppliers of manpower, even if it happened only once. As a reminder, Cassius Dio told us of 'eight thousand cavalry, fifty-five hundred of whom he sent to Britain.' But the question of who those cavalry *were* needs to be examined.

The Sarmatians were a confederation of tribes from the steppe with Iranian links, including the Roxolani, the Iazyges, and the Alani. Herodotus tells us their women were to be found 'frequently hunting on horseback with their husbands; in war taking the field; and wearing the very same dress as the men.' This is supported by the finding of many grave mounds that contain the bodies of female warriors, accompanied beyond death by their weapons. So when Cassius Dio tells us that 5,500 cavalry came to Britain, we don't have to imagine that they brought another 5,000 or so family with them, for it could be that even

Reenactor in the role of a Sarmatian woman. Note the long ring-pommel sword, the recurve bow, and the lack of a Roman four-horned saddle. (*Author's collection*)

half that number were women. Rome, of course, had no provision for female soldiers, and so if this was the case, then only half that number of riders would actually be co-opted into the *auxilia*, which might account for why we have not found more evidence for their activity on the island.

Given the unusual nature of the Sarmatians' recruitment, it should come as no surprise that on the only two inscriptions that bear names, those names are Roman. One is of a prefect, Aelius Antoninus, and such an officer, of course, would be an equestrian citizen anyway. The names on the other source, though, (*RIB* 594) refer to a rider who was a bodyguard, *a singularis consularis*, and his family. The rider is named as Julius Maximus, and with the stone he commemorates his wife Aelia Matrona, her mother Campania Dubiata, and his son Marcus Julius Maximus. It is possible that upon their transfer to the *auxilia* and removal to Britain, the 5,500 Sarmatians were made to take up Roman names, not only allowing them to fit easier into a Roman mould, but also to distance them a little from their origins. It may instead be that this one memorial dates from a later period, when the Sarmatian unit had been largely Romanized, and that future finds will reveal solidly Sarmatian names in Britain.

The Sarmatians were great horse people, and their steeds were larger and more powerful than the small Roman ponies. It seems likely that the introduction of their horses into Britain forms the origin of the large breeds on the island, and their breeding of the animals in the northwest spread across Britain over the following centuries. One fascinating find at Ribchester is what appears to be a '*tamga*'. A *tamga* is a mysterious and complex symbol associated with the varied horse peoples of the steppe, from the Scythians to the Mongols, often believed to have been used for branding animals. This item, then, is a *very* non-Roman item, found at a Roman fort and connected to the Sarmatians' earliest origins.

On Trajan's Column, Sarmatians are shown in the form of armoured cataphracts (something we shall examine with their equipment later), and so they may have supplied these heavily armoured units for the *auxilia* in Britain. As such, as well as their acknowledged presence at Ribchester, and possibly Catterick, they may also be equated with the Equites Catafractariorum placed at the unidentified fort of Morbio in the *Notitia Dignitatum*. Again, it is worth noting that despite their geographic separation, the Sarmatians and the Parthians, the two people who seem to have developed the cataphract, both descend from Iranian stock and share deep-rooted links.

The cultural contribution of the Sarmatians to Britain is strong. Our fascination with these people only deepens, and the repeated attempts over the decades to link them to the legend of King Arthur and to Lucius *Artorius* Castus of Dalmatia (*CIL* XXX, 01919) only goes to show the hold they have on the British imagination.

Across the Danube from the Sarmatians, within the empire, lay Thracia, a region attested as having supplied a large number of men for Rome. As far back as the civil war with Pompey, Caesar tells us '[King] Cotys had provided the same number [of men] from Thrace and had sent his son Sadala.'[4] He goes on: 'The camp was being zealously defended [...] much more keenly still by the Thracians and barbaric auxiliaries.'[5] At this stage, Thracia was a client

4. Caesar, *Civil War* 3.4.
5. Caesar, *Civil War* 3.95.

kingdom of Rome, and remained so until the mid-first century, when it was incorporated as a province.

That Rome valued the martial strength of the Thracians is suggested also by the fact that one of the most celebrated classes of gladiator in the arena was known as the Thraex, armed with a distinctive curved blade of Thracian origin called a *sica*. In fact, the very martial nature of the Thracians is also suggested by an inscription (*AE* 1944, 80) which tells us that Aurelius Dionysodorus recruited Thracian brigands into the Roman *auxilia*!

The Thracians were a distinctive people culturally linked to the Dacians, from a region that covered roughly modern Bulgaria and European Turkey, and by the time of their submission to Rome they had been influenced heavily by both Greek and Macedonian cultures. That they retained their own ethnic identity in the *auxilia* more than many of the other cultures is suggested by the fact that far more Thracian recruits returned to their homeland after their term of service than other peoples, who often settled in their area of service upon retirement. The recording of distinctly Thracian, non-Roman names in epigraphy also suggests a retention of their culture throughout service. Such figures as Longinus Sdape, son of Zematucus, buried at Colchester (*RIB* 201), or Fuscus Dorilsis, son of Eptaecenti at Astorga (*AE* 1928, 165), both from Serdica, are prime examples.

How the Thracian units in the *auxilia* must have cheered on the accession of Emperor Maximinus Thrax in AD 235, for he was more or less one of them. Maximinus has the distinction of being the first truly low-born provincial who rose to the purple. Reputedly a shepherd in his boyhood, the brutal giant is stated in several sources to have served as a rider in an auxiliary cavalry unit. To an ordinary soldier on the fringe of empire, Maximinus must have struck a chord, since if he could manage such a thing, then perhaps any Thracian auxiliary could set his sights on high office, something that becomes increasingly true and possible as the next century rolls on. Maximinus was said to have had two barbarian parents, and is claimed by Ammian to have descended from the Carpi, a people who have ancestral links in both Thracia and Dacia. It is possible that this is why the only unit in Britain known to have taken his name as an honorific was a Dacian unit: Cohors I Dacorum in Birdoswald.

The Dacians, within the empire yet on the northern side of the Danube, are another people whose culture may have remained apparent in the units despite their removal from their ancestral lands. Indeed, in Cohors I Dacorum, we see not only an example of their cultural heritage transplanted from the Balkans to Northern Britain, but also how such cultural signs survived down the years long after the unit stopped receiving recruits from home. Two of the inscriptions found at Birdoswald bear carvings of a *falx*, the distinctive curved Dacian blade that was so brutally effective against the Romans in the Dacian Wars that the cross-reinforcing of Roman helmets to strengthen against such a weapon became the norm. These two inscriptions (*RIB* 1909 and *RIB* 1914) both name the unit, both show the *falx* clearly and both date to the early third century, some eighty or ninety years after their raising under Hadrian. Since they would be on probably their fourth generation of recruits by this time, most of whom would have come from local areas, or at least from northwestern Europe, it is highly unlikely any of the cohort had ever even *seen* a *falx* or laid eyes upon Dacia, yet they still proudly display the blade in their work. Indeed, no evidence of the existence of an actual *falx* has been unearthed in Britain, yet the carvings are a solid representation of the weapon.

It is also interesting to note that *RIB* 1909 commemorates the construction of a granary by the men of both I Dacorum and Cohors I Thracum, two neighbouring lands, whose traditional weapons (the *falx* and the *sica*) bear a marked resemblance in shape and design, if not in size.

Thracian *sica* and Dacian *falx*. For the image of the falx mentioned above, see the entry in the listing for Cohors I Dacorum. (*Wikimedia/Petter Bøckman & Wikimedia/Rabax63*)

Dacian cultural signs in Britain crop up also in names. An undated inscription from Birdoswald (*RIB* 1920) was set up by an unknown character to his brothers Blaesus and Decibalus. Blaesus is a solidly Roman name, attested long before Dacian recruitment, while Decibalus is a staunchly Dacian name. (Decebalus having been the Dacian king whom Trajan fought at the start of the second century.) As such, it is possible that this is an earlier stone, at a time when truly Dacian recruits were still extant in the unit. Interestingly, another Decibalus (probably not the same one) scratched his name into a bowl (*RIB* 2501.156) at Chesters 20 miles away, a fort with no other Dacian connection.

A little closer to home than Dacia, and a lot closer to Rome, lay the land of Illyria, Roman since AD 27 as the province of Illyricum. The coastal region of Illyricum, facing Italy across the Adriatic, is known as Dalmatia, and is another supplier of auxiliary troops to Britain. Unfortunately for those of us seeking to examine cultural retention, the peoples of that region had been so strongly influenced by Greek culture and then Roman that by the time they were supplying auxiliaries to Britain there was hardly anything outstandingly Dalmatian about them, visually speaking, to look for. Though we can see no native Dalmatian names arising in British records, Cohors IV Delmatarum records men by the names of Beusas, Daverzeus and Bato in their earlier posting at Bingen, all solidly Illyrian in nature.

We have similar issues with Pannonian recruits, who have left us nothing eye-catchingly native to look for. No names with Pannonian/Illyrian roots survive in Britain, though one Balaesus (a Pannonian name) is commemorated on a stone in Portugal. The enigmatic 'Dagvalda' from the stone at Cawfields milecastle, while he served in a Pannonian unit, was probably Germanic, even Batavian, given the similarities between his name and other Batavians. As such, the available evidence of Pannonian culture in Britain is ephemeral at best.

Our troubles with cultural anonymity only increase when we look to Germanic and Gallic units. The most notable cultural detail in this era is that which we have already examined: the Batavians' skill in crossing water. Not only had Roman, Gallic and Germanic cultures been bleeding into one another for some time before the creation of a standing *auxilia*, making identification difficult, but in terms of equipment we have a further problem. As we will see shortly, Rome adopted the cultural armament of the Germans to some small extent, and the Gauls to a much greater one, making the differences between the peoples in visual terms negligible. It is to some extent possible to see Germanic culture surviving in odd equipment fittings and decoration, but again it is difficult to say how much bleed there was between Germanic units and other parts of the army and the regions in which they served, and so it is far from impossible that anything which appears to be culturally Germanic or Gallic is actually generic Roman by the time of the mid-Principate.

One thing that seems to have been a Gallic and Germanic cultural practice was the taking of heads in battle. There is plentiful evidence of this, and Diodorus gives us evidence of the practice among native Gauls:

> The Gauls cut off the heads of their enemies slain in battle and fasten them about the necks of their horses. They hand over the blood-stained spoils to their attendants to carry off as booty, while striking up a paean over them and singing a hymn of victory.[6]

6. Diodorus, *Bibliotheca Historica* 5.29.4.

A quick glance at the tombstone of Insus of the Treveri, an auxiliary cavalryman serving in Lancaster clearly shows him on horseback, carrying a severed head by the hair, a brutally clear piece of evidence of Gallo-German cultural practices continuing into their service to Rome.

Again, the Iberian tribes, having given Rome the 'Spanish sword', and having been an integral part of the Roman military machine for over a century before the creation of the standing *auxilia*, have left few cultural footprints for us to follow, barring a little individuality of their cavalry horse trappings. Notably, Iberia supplied *auxilia* until the reign of Nero, having been a troublesome, restive people for much of their previous Roman history. From the end of the Julio-Claudians onwards, though, Iberia became a very settled and thoroughly Romanized province, likely why it stopped supplying men to the *auxilia* in force.

It should now be apparent that ethnicity in the *auxilia* seems to fade with time, as recruiting became more local, and native blood thinned in the ranks, yet certain aspects became, if anything, *more* important as regimental symbols. As a brief aside, it is worth noting how certain provinces and regions attempted to retain a hint of their pre-Roman culture, even in military imagery.

As these *auxilia* slowly became more Roman, and Rome in return became more and more multicultural, we see the *auxilia* shifting from what began as an almost barbaric, peripheral status, the outsiders of the Roman military,

Carving of Dalmatian arms and armour from Split, dated to the late first century BC or early first century AD, the style archaic and non-Roman, Split Archaeological Museum. (*Author's collection*)

to something much more central and entwined with it. Increasingly formed of citizens, ever more homogenous, they became involved with the centre of Roman power; the imperial German bodyguard is a prime example, while the notion that Ala I Hispanorum Asturum gained their 'Praetoria' epithet for a similar position with the Severans is another. The ultimate expression of ethnicity becoming less important over time can be seen in the increasing instances of emperors rising from the low-born in the provinces. As well as Maximinus Thrax, ex-auxiliary cavalryman from Thracia who we met above, examples include Macrinus from Mauretania, Aurelian possibly from Dacia, Probus from Pannonia, Diocletian from Dalmatia, and a bunch of Severans from Syria. In fact, the area of Sirmium in Pannonia produced no less than ten emperors. Indeed, Pertinax, himself of Gaulish blood, apparently commanded a Tungrian cohort in Britain (information courtesy of the Brühl inscription and the *Historia Augusta*). That the emperor of Rome commanded an auxiliary unit in Britain is impressive; that he was of Gallic blood, leading Germans against Britons in the service of Rome is truly fascinating.

Having touched upon it with respect to inscriptions and names, one aspect of the ethnic makeup of the auxiliary units that deserves a little investigation is language. To some extent, language is a more difficult yardstick for ethnicity than archaeological remains, for a few very simple reasons. Rome was a largely bilingual state. Though dozens of languages would persist across the empire, official, religious and military matters were generally dealt with in the West in Latin, and in the East in Greek. These two tongues were, between them, the lingua franca of the empire. As such, it matters not what the ordinary soldier spoke when alone with his mates, anything he left in official terms was generally in Latin or Greek, and so masks any ethnicity. Even the use of Greek seems to have been at times more tolerated than accepted. We find a story in Suetonius in which Emperor Claudius 'not only struck from the list of jurors a man of high birth, a leading citizen of the province of Greece, because he did not know Latin, but even deprived him of the rights of citizenship.'[7] This might be a lone case, but it does suggest a trend. On the other hand, Suetonius also has Claudius use the phrase 'Since you are ready with both our tongues.'[8]

Whatever prejudices may have existed, it is abundantly clear from surviving inscriptions and documentation that Greek was the standard in the East. Latin could spread west among a group of peoples, bringing them together to some extent, but that had already happened in the East with the spread of the Greek world and then the Macedonians. As such, Greek was already so widely spoken,

7. Suetonius, *Claudius* 16.2.
8. Suetonius, *Claudius* 42.1.

Latin was not going to replace it easily. Both languages, for Rome, formed the core of empire-wide communication and record keeping, in much the same way as English has become a world standard and remains as much thanks to the need for standardization in a world run by computers. As such, we can see Latin and Greek as imposed upon all the cultures of the empire for simple expediency (and almost certainly a little imperious smugness).

Remarkably, for the era, Rome appears to have managed a high level of literacy among its armies. It was, of course, the first truly bureaucratic military, and the sheer level of paperwork would require very literate clerks and officers (we have very little evidence of illiterate officers), but also of the ordinary soldiery. If nothing else, a soldier was going to want to know what his diploma said. The various sites that have turned up surviving parchment or tablet records have given us a glimpse into the ordinary world of the auxiliary, and if those soldiers could not write, we would not have this source. Moreover, the vast quantity of phrases and slogans, as well as names, graffitied on items across the empire tells us just how ready the ordinary soldier was to write:

> OMO[…] LIINDI[…] LEGE FEL[…] – 'Omobonus of Lindinis read this happily' – (*RIB* 2491.155).

Or the quarry inscription:

> APRO ET MAXIMO CONSULIBUS OFFICINA MERCATI – 'In the consulship of Aper and Maximus the work-face of Mercatius' – (*RIB* 1009).

Or perhaps the best of all, the recent find at Vindolanda of a stone carved with a phallus that also bears the slogan 'SECUNDINUS CACOR', or Secundinus the 'shitter'!

This broad use of Latin, of course, helps account for the lack of linguistic evidence for Germanic, Gallic, Sarmatian, Mauri people and other such peripheral groups. These cultures had not developed a written language, and so the soldiers left us any records in the lingua franca (an ironic phrase, when we are talking about the people of Gaul having no written tongue). The army would undoubtedly need a good level of Latin or Greek for the understanding and following of commands, although Arrian does suggest that orders were sometimes given in native tongues: 'Celtic to the Celts, Getic to the Getae, and Raetian to the Raetians.'[9] Since Arrian wrote in the second century and is at that point in his treatise speaking about cavalry, it is possible he is referring to the more barbaric

9. Arrian, *Tactica* 44.1.

numeri rather than standard *auxilia*. Certainly, it seems likely that irregular units might react easier in their own tongue, and the potential headaches of telling a prefect from comfortable Umbria that he needs to learn Getic for his next command seems to make the use of native languages for military commands troublesome at best.

Cast of the tombstone of Hermes of Commagene in Brough church. (*Author's collection*)

There is, however, evidence of native language use in the provinces, despite their being a lingua franca. That Greek appears in Britain should not be a surprise when units from the East were posted here, but the finds of Greek, interestingly, do not entirely tally with where Greek-speaking units might have been based.

One of the most fascinating inscriptions in the north of England was found at Brough-under-Stainmore (*RIB* 758), the third-century epitaph to the 16-year-old Hermes of Commagene. We have no record of a garrison at Brough until the rather vague Numerus Directorum in the *Notitia*, though among the many lead seals found there, two belong to Thracian units.

Corbridge has supplied two altars in Greek (*RIB* 1124 to Astarte and 1129 to Heracles of Tyre) where no potential Greek-speaking units are attested. Perhaps one of the most interesting finds comes from the *mithraeum* of Rudchester (*RIB* 1397). This altar was written in the Latin alphabet, but stood in the temple of an originally Persian deity, celebrated the god Apollo, with the Greek word for invincible transliterated into Latin, and all in a fort full of Germans! Another fascinating contextually difficult find is *RIB* 436, a gold amulet written in Greek and bearing a magical prayer for protection to the Lord God of Hosts, Yahweh. The amulet was found at Caernarvon, a fort whose only known garrison was German. Similar amulets have been found at York and Torksey.

When added to the fact that the Hercules altar in Corbridge was raised by a priestess with a Greek name, an interesting possibility occurs. The Roman priesthood was a pragmatic thing, more a political appointment than anything to do with faith (certain specific cults notwithstanding – Cybele, Vesta, Mithras etc.) There seems to have been no provision in the army for priests, with officers leading important rituals, as they would have done back in Rome as higher-status civilians. Parts of the empire, though, had long traditions of dedicated priesthoods, and might not be comfortable, or even able, to put matters of faith in the hands of a non-priest. As such, that might explain soldiers having local priestesses raise altars. And if the soldiers we are talking about were, for instance, of Germanic origin, they would have no written language of their own, and it would be the priest or priestess who would inscribe the altar or amulet. Given that the bulk of such priesthoods came from the eastern provinces, perhaps it is not so odd to find Greek-speakers doing the work.

Two other pieces of written evidence leap to mind as displaying an ethnicity that stands out. The first is the tombstone of Regina that we have already encountered at South Shields. The main three lines of the stone are in Latin and in a standard format, telling us 'To the spirits of the departed and to Regina, his freedwoman and wife, a Catuvellaunian by tribe aged 30, Barates of Palmyra set this up.' Below this very Roman inscription, though, is an extra line, outside the main box, and carved in Aramaic, the language of Palmyra. Where the Latin is very formal, the native script rather touchingly simply says: 'Regina, the freedwoman of Barates, alas', presumably the direct words of her husband in his native tongue. If Claudius could strip a Greek of his citizenship for not speaking Latin, I wonder what he would have made of Barates.

And finally, though tombstone *RIB* 2182 from Auchendavy is in good Latin, it is interesting to note that both the deceased and the person who raised the stone were named Salmanes, a Semitic name. One might assign this grave, as well as the Jewish amulet from Caernarvon to Jewish settlers fleeing their homeland in the wake of Hadrian's brutal punishments, rather than suggesting they have any bearing on the *auxilia*, though as examples of ethnic diversity they are still worthy of note. Indeed, at Intercisa, the Jewish population became numerous enough to construct a synagogue.

Perhaps following on from language it is appropriate to have a look at religion. While it is the Hamians, the Dacians, and the Sarmatians who have left us the most striking visual reminders of their culture, and the Gauls and Germans have been hard to pin down, when it comes to gods, we find much the reverse, though the reasons for this are clear.

The vast majority of the altars found in Britain could be said to be either standard Roman or local British Celtic, with those inscribed IOM (Iupiter Optimus Maximus – Jupiter Greatest and Best) topping the list. But it should be noted here that many of the Eastern gods and goddesses had already been embraced by Rome that there was nothing unusual in their worship anywhere in the empire. The most striking example of this, of course, is the soldiers' religion:

Mithraism. Mithras was a Persian deity that became so tied to the Roman military that his temples are evident at many forts across the province and beyond, as well as in cities.

Similarly, Isis had already become to some extent a 'Roman' goddess even before the invasion of Britain. Serapis had come from Egypt to become embedded in Roman culture, and similarly Sol Invictus from Emesa.

Relief from Housesteads fort illustrating the birth of Mithras from an egg, surrounded by signs of the zodiac. Though by this time solidly Roman, it is hard to ignore the exotic and oriental flavour of the carving. Now in the Hancock Museum. (*Author's collection*)

Though Dolichenus is a term derived from the city of Doliche in Anatolia, close to the Syrian border, Jupiter Dolichenus is clearly a god either created or appropriated by Rome, and so we might say that the in-excess-of-a-score of altars to that god found in Britain are no more ethnic than Apollo, himself a Greek god. That the Greek gods had already become identified with the Roman pantheon means that many Eastern peoples would not need to stray to Greek names, for they could comfortably use the Latin alternatives. This combination makes it easy to see why there are not many gods attested

Sol Invictus carving in Corbridge Museum. Note the similarity in the carving of Sol from Corbridge to the carving of the Palmyrene deity now in the Louvre. (*Author's collection*)

in Britain from the East and South. Herakles and Astarte, as mentioned above at Corbridge, are joined only by Hammia at Carvoran and by 'Syria', a rather vague characterization of the province and region unless it is to be read as Suria (see below), to a grand total of three culturally 'Eastern' gods in Britain. In the appendices you will find a list of the deities attested on altars found in Britain, but to summarize the numbers: four from the East, forty-eight broadly Roman, and twenty-nine that appear to be Celtic (it is common to find some of these gods worshipped on both sides of the English Channel, and so to attempt to divide them between British, Gallic and Celtiberian would be fraught with difficulty).

The one other group that stands out is the Germanic deities. While the British, the Gauls, and swathes of Iberia shared a culture that had seen their gods proliferate and spread, the Germanic peoples belonged to a different tradition, more in keeping with Scandinavia than with the Celts. As such, their gods stand out. Ten German gods appear on British altars, compared to just four from the whole East. Above this number we also find an altar from Castlesteads (*RIB* 1989) bearing the inscription 'To the Mother Goddesses beyond the seas', and, given that of the four units associated with the fort two were Gallic and two German, it seems likely that this, too, was raised to German gods, much like *RIB* 2064: 'To the German Mother Goddesses.'

There is a certain amount of uncertainty with the gods of Roman Britain, for some, although they appear to be Celtic, could be Germanic, and so it is possible that Arciarcon and Garmangabi are also German. Similarly, there is ongoing debate about whether the altars at Carvoran and Catterick (*RIB* 1792 and 726) do indeed refer to Syria, or to some Celtic goddess by the name of Suria, which would drop the Eastern gods represented on the island to a mere three.

Essentially, with the exception of a number of Germanic gods specifically honoured, and a smattering of Eastern ones, the gods of Roman Britain were widely enough worshipped that it is difficult to pin any sort of ethnicity to them. The very epitome of this comes in the form of *RIB* 653 from York, an altar raised by a soldier of Legio VI to 'the African, Italian, and Gallic Mother Goddesses.' In far off Dacia, at the fort of Micia, multiculturalism shines through in that two of the fort's garrisons went so far as to build temples to their native gods, Syrian archers to Jupiter Turmasgades, and the Numerus Maurorum to their African gods.

In summation, it is readily visible how complex a subject the ethnicity of the *auxilia* was, but some cultures seem to have been represented longer by their regiments than others, and there was a general decline in any form of ethnicity over time. It seems possible that the *numeri* were brought about by the need to infuse the army with a fresh wave of barbaric violence since the *auxilia* had become part of the general establishment. As such it is tempting to see the

Two images of the Egyptian god Horus, thoroughly Romanised, both on display at Wallsend. (*Author's collection*)

numeri even from the second century onwards, as far more ethnic than the rest of the *auxilia*.

One thing it is hard to detail is any ethnicity displayed by the Britons themselves against the Roman invader, although in other provinces they likely blend in in the same manner as the Gauls did here. One thing that Cheesman notes is the distinct lack of officers from Britain among auxiliary units, and raises the suspicion that the Britons were too proud a warrior people to want the career path such a position would offer. While this appeals to me, as a Brit myself, it seems to be reaching. The Britons were not greatly removed from the Gauls and the Germans, who both provided officers, and men like Cogidubnus tend to put the Britons on a slightly less noble and more selfish level.

We shall leave the subject of ethnicity of units with a wonderful example. *CIL* VIII, 9381 (a tombstone) from Cherchell in Algeria tells us:

> *Dis Manibus sacrum Sextus Iulius Iulianus ex Germania superiorem tribunus numeri Syrorum Malvensium hic sepultus est dum deducit iuniores Bessos mille in Tingitanam provinciam.*

Which translates as:

> To the spirits and the sacred memory of Sextus Julius Iulianus from Upper Germany, tribune of Numerus Syrorum Malvensium was buried here while he was leading a thousand Bessi to the province of Tingitana.

What this tells us, in short, it that a German, who was the commander of a Syrian unit named after a Dacian town, led 1,000 Thracians to Mauretania!

Chapter 4

Fabrica

(Arming the forces)

The subject of Roman military equipment fills shelves of books, and our knowledge changes constantly with every new find. For anyone with an interest in studying the subject in depth, I would point them at Bishop and Coulston's *Roman Military Equipment* as a starting (and main) point. While we are here only really interested in auxiliaries, it is impossible to look at the *auxilia* without at least peripherally studying their legionary counterparts, for it is sometimes hard to distinguish the two.

Until the advances made through archaeology in the twentieth century, the majority of our knowledge of Roman arms and armour came from sculpture. Such sources come in several forms, of course, the two main ones being great state monuments such as Trajan's Column, and more individual ones like tombstones. The best visual sources are usually the great monuments, and even now, the majority of the public see the Roman military as it is displayed upon Trajan's Column, even re-enactors basing their kit upon its images. There is a problem with this, though. Such monuments were made to serve a purpose. Trajan's Column was not designed as an accurate record of the Dacian Wars, but as a piece of propaganda. As such, it displays the Roman army in an idealized fashion. It emphasizes the foreignness of some of the forces on its images, and goes to great pains to separate out the citizen soldiers from the *auxilia*. That legionaries are depicted almost identically throughout every scene, and the same true of the *auxilia*, should be a warning sign. No military force, even now, manages such conformity, and that suggests that the designers and artisans of the column were not even attempting to represent the forces accurately.

Of course there is a clear reason for this. The column was constructed in Rome, by artists and workers in the city. It is extraordinarily doubtful that many of them had ever seen the legions and *auxilia* in action in the provinces. Their portrayals would be based only on what they knew, and what they knew were the units stationed in Rome. Thus the legions on the column are likely based upon the praetorian cohorts and cavalry, with a few extra details thrown in, while the *auxilia* were probably a cobbled together image from other sources.

Rome would have seen marines of the fleet, for they were based in the city, working with the Colosseum's sunshade if nothing else. We are still uncertain of the usage of the building on the Caelian Hill named the Castra Peregrina, but its name (camp of the foreigners) suggests its use to house non-citizen troops when in the city. If this building was regularly occupied by auxiliaries, inspiration for the column's soldiers might have come from there. But still, even with models in the city, the sources for the artisans were scarce. As such, we can hardly conclude that the imagery on the column is an accurate representation of the Roman army. The same can be said for the Column of Marcus Aurelius, the Cancelleria reliefs, the Great Trajanic Frieze, and all other monuments in the city that show the Roman military. They all had a purpose, and that purpose was never accuracy. This is not to say that they are complete flights of fancy, of course. Clearly there is a hefty element of truth in the carvings. We just cannot rely upon them without question.

There is an alternative imperial source, however. Some monuments have a more trustworthy background. The best surviving example is the Tropaeum Traiani at Adamklissi in Romania, celebrating the conclusion of the Dacian Wars. Its series of carved metopes, most of which survive in varying conditions, also portray military scenes from the Trajanic era. Given the location and circumstances of this monument, it was almost certainly carved by the soldiers in situ, or at the very least by locals who had a good working knowledge of their subject. As such, those carvings, cruder than the great monuments of Rome as they are, constitute a much more trustworthy source. Other provincial sources exist, too, of course. The Column of the Boatmen in Paris, the Mainz pedestals, the

Imperial Roman monuments (L-R: Trajan's Column, Column of Marcus Aurelius, Tropaeum Traiani). (*Author's collection*)

distance slabs of the two walls in Britain, and various other provincial carvings would have been made by soldiers or at least by those familiar with the soldiers.

Private tombstones, however, give us the best source (not that we should always take such images entirely at face value, as I'm sure not every cavalryman spent his time stabbing down at cowering barbarians). But while tombstones were designed to laud the subject, there was no agenda involved that required altering the details of equipment and uniform, and so we can be content that there is a lot more truth to such sources.

Archaeology, of course, gives us the best record of equipment, and it is largely by comparison of existing imagery with archaeological finds that a better image of Roman military equipment is slowly being pulled together. To this we can add written records, including some snippets from ancient writers, but perhaps more importantly finds of tablets and papyri at sites around the world. Graham Sumner has done tremendous work on the subject of uniform, design, materials and dyes, largely from the study of such sources as papyrus records and the Fayuum mummy portraits.

The upshot, then, is that little can be said with absolute certainty, and what we consider to be the most likely explanation for any military equipment or garb might change with the next find. This problem is, of course, compounded by time. In studying the *auxilia* in Britain alone, we are investigating four centuries of an institution, and clearly there would be plenty of changes over time, so we

Soldiers on Trajan's Column, idealised, with legions on the left and *auxilia* on the right. (*Author's collection*)

Two second century bronze statuettes in the British museum. As with Trajan's Column, identifiable by their armour. (*Author's collection*)

The tombstones of the Annaius Daverzus (Coh IV Delmatarum cordus) and Flavoleius Cordus of Legio XIV Gemina, identifiable only by their weapons. (*Images: Wikimedia/ JordiCuber and author's collection*)

cannot build a single picture of an armed and armoured auxiliary and presume he would be the same in AD 410 as he was in AD 43.

A second issue is the identification of who is actually an auxiliary among military depictions. A similar problem has plagued the study of the praetorians, who are hard to identify among Rome's soldiery. We can pick out the cavalry by the simple fact that in imagery they are almost always on a horse, and in archaeology they will be found in the same context as horse equipment. The difference, however, between a legionary and an auxiliary infantryman can be very difficult to tell. On gravestones, for instance, when soldiers are often shown in their tunics without armour, it can sometimes be only the attached inscription that gives us a clue. Otherwise they could easily be either.

Pages 82 and 83 show three examples for comparison, illustrating the difficulties.

Now that we know we will be plagued with doubt, let's turn first to the basics. A good starting point for Roman military equipment is the statue known as the 'Warrior of Vacheres'. This statue, carved in the reign of Augustus and found in France, has as its subject either a Roman or a Gaul, and therein lies our point of interest. The chain shirt the man wears appears to be a Roman *lorica hamata*, with shoulder doubling. His cloak has a Roman feel to it, and the shield would look at home on any Roman military carving. Yet he wears a Gallic torc around his neck, he seems to be wearing a sleeved tunic, and possibly trousers, and what can be seen of his sword hilt looks less Roman than one might expect. Not knowing who the man was, or the purpose of the statue, makes it difficult to call judgment on any of the equipment, but the simple fact that it is difficult to tell whether this is a Gaul with Roman-style items or a Roman with Gallic-style items is extremely telling.

The Warrior of Vacheres in Musee Pierre Martel. (*Author's collection*)

In the same way that Rome owed a huge cultural debt to Greece, so she owed a martial debt to Gaul. In early Imperial Rome the Celts were both feared enemy and fascinating neighbour, representing the 'noble savage'. Tacitus delves heavily into this in his *Agricola*, and Caesar treats his opponents as such in his *De Bello Gallico*. Rome had spent centuries at periodic war with the Gauls, while culturally appropriating aspects of their world, long before Caesar's use of Gallic and Germanic cavalry. By the end of the Republic, the Gauls were the only people who had ever sacked Rome, an event that still simmered in the Roman consciousness. They could not help but appreciate the martial abilities of the Gauls, even while cursing them. In this light, that Rome might move towards Gallic models is hardly surprising.

In his *On the Latin Language*, Varro tells us of the *lorica* as a 'Gallic corselet made of iron', describing it as a 'shirt made of links'.[1] Clearly, then, the chain shirt (I shy away from the word 'mail', being a medieval French term) was adopted by Rome from the Gauls. Similarly the Roman cavalry sword, the *spatha* as it is now known, can trace its roots to the Gallic/Celtic long sword, which the Romans initially scoffed at for being too long and heavy and requiring a wide swing. Such a weapon did not suit the Roman legions' way of fighting at the time, and yet it did suit cavalry, as the Gauls had shown to great effect. It is also possible to trace the development of Roman helmets, from the time of Augustus onwards, moving away from the Italian Montefortino models and towards new designs heavily based on Gallic models. The Roman saddle is a Gallic invention. Later in the empire, Gallic breeches were adopted. Gallic shield shapes and designs… the debt goes on.

Indeed, there are relatively few aspects of Roman equipment that do not owe a debt to the Gauls. The gladius, of course, known as the 'Spanish sword' is one, and yet the name makes it clear that this is Rome owing a martial debt to *another* neighbour. The spear (often referred to as a *hasta*, though with a number of names for variants in design) is a multicultural weapon.

Roman sources suggest that the legions and the *auxilia* were equipped differently, with Tacitus saying:

> The soldiers immediately seized arms without regard to military custom or rank, with no desire to distinguish praetorian or legionary by their proper insignia; they wore the helmets and shields of auxiliaries without distinction.[2]

This, however, only tells us that legionaries grabbed the gear of the auxiliaries. It may *imply* that this gear is different, but that is not a given. Indeed, that he notes the only way to distinguish between legionary and praetorian kit was by the insignia lends weight to the notion that there was actually no set distinction across the military. The main source usually quoted to claim a difference

1. Varro, *De Lingua Latina Libri* 5.116.
2. Tacitus, *Histories* 1:38.

between legion and *auxilia* is Trajan's Column, and we have already noted the difficulties with this.

For infantry equipment, the debate of armour has raged for decades and continues to do so. Quite ignoring the hot topic of whether such a thing as 'leather armour' ever existed, there is a debate over the usage of plate armour. We know that chain was favoured by both *auxilia* and legions. It does not have the same frontline protection as plate, but is more flexible, and is displayed on gravestones of both *auxilia* and legionaries. The same can be said for scale armour. The famous '*lorica segmentata*', however, remains divisive. If we were to trust Trajan's Column, we would say that segmented plate is solely the province of the legions, though archaeology suggests something else. Given the sheer scale of *segmentata* finds at auxiliary forts, it is hard to deny the possibility that the *auxilia* also used this heavy armour. The best minds in the field remain undecided. Though no auxiliary *gravestone* shows *segmentata*, neither do legionary ones. Indeed, *segmentata* appears only on friezes and monuments, and even then not all that often. That the heads of *pila* (a legionary weapon) have also been found at such sites might suggest a common legionary presence at auxiliary forts, which further muddies the water.

The heaviest armour, the *segmentata* provided an immense protection, particularly on the shoulders, where blows in the front line would likely land when glancing off a helmet. The armour consists of numerous curved steel plates

Lorica segmentata discovered at Corbridge, with partial reconstruction, Corbridge Museum. (*Author's collection*)

that encase the torso, with flat plates on chest and upper back, and articulated shoulder plates, all fastened together inside with leather straps, and hinged at critical joints. The armour would likely have been worn with a 'subarmalis', a leather, wool or linen waistcoat that took some of the weight, prevented chafing, but also acted to absorb some of the shock of blows.

The squamata, or scale armour, is the least common in both finds and representations, though its appearance on monuments such as Trajan's Column, on some tombstones, and fragments found at forts, suggest that it was far from rare. The armour consists of metal scales sewn overlapping in tight rows onto a shirt of padded linen, or possibly leather. This gives a good mix of flexibility and protection, but would be expensive and time consuming to construct, which might be a reason that it is most often depicted on centurions, standard bearers and other higher-ranking soldiers.

By far the most common of all Roman armour is the chain shirt. The construction of chain is well-known, formed of many connected iron links. Chain armour provides good protection against penetrative strikes, but does not rob a blow of its strength in the same manner as *segmentata*, so it is possible to break bones through it without breaking the links. This may be why chain shirts initially had a doubled thickness around the shoulders, backed with leather over the main shirt and fastened with ornate bronzed hooks, although the shoulder-doubling seems to go out of use are the end of the first century. Chain shirts were worn to mid-thigh and were generally short sleeved.

It is possible that segmented armour was reserved for the front lines in any fight, given the danger there of taking blows on the shoulders. Having trained in re-enactment, I can attest to the extreme challenges of throwing *pila* while wearing *segmentata*. The throw often goes wild as the plates rack up and jam into the bicep, and the results also include wicked bruising from elbow to shoulder. Indeed, on the Lambaesis inscriptions (*ILS* 2487), Hadrian says as much: 'You have done the most difficult of all difficult things; you have thrown the javelin clad in armour.' As such, I see *segmentata* being reserved for the front ranks, the rest in chain, with which they could launch *pila* over their comrades into the enemy. Despite the persistent notion that the *auxilia* were fielded as light infantry and the legions as heavy, evidence suggests there was no such distinction in practice, with Tacitus telling us of Mons Graupius: 'Agricola encouraged three Batavian and two Tungrian cohorts to bring matters to the decision of close fighting with swords.'[3] As such, it is far from impossible that the front lines of auxiliary cohorts in battle also wore *segmentata*, which might account for the finds at so many of their forts. The truth of this still eludes us

3. Tacitus, *Agricola* 36.

and will continue to do so until new finds shed any further light. Barring the possibility of *segmentata*, though, current confirmable evidence points to the *auxilia* commonly using shirts of scale or chain.

In terms of helmets, the prevailing view is that auxiliary helmets were less ornate and of cheaper construction than those of legionaries, although there is actually no evidence of this. Very likely, given the huge variety of helmets uncovered across the empire, with so many different types in different materials, soldiers bought the best helmet they could afford, regardless of its style and their military status. The exception to this would be cavalry helmets, discussed shortly. Even Trajan's Column, which works hard to separate out the legions and the *auxilia*, gives them the same helmet. There are occasionally arguments that ethnic units might adapt their uniform to more familiar designs. While there is no direct evidence for this, it cannot be easily refuted. The infamous 'Batavian helmet' is an interesting item. It consists of a Roman helmet that has been stripped of extraneous parts and added to with feathers and leather. It seems to date from the Civilis Revolt, and probably signals an enemy of Rome stamping his native identity upon it, but it does make one wonder whether units might try such a thing in service to a lesser extent.

What we *can* say about the 'standard' auxiliary infantry's equipment is that it is different from that of a legionary, principally in the form of weapons. In grave stele images, legionaries are armed with gladius, pilum and large, curved body shield. By comparison, auxiliaries are uniformly portrayed with a *hasta* – a long spear, rather than a javelin – and usually a pair of them. The auxiliary hasta was an eight-foot ash shaft with a narrow, leaf-bladed iron head, used for stabbing rather than throwing, the use of which clearly gives the auxiliary infantry different possible roles and tactics in battle to the javelin-casting legions. To complement this, they carried a flat shield, oval in shape. Their armoury was completed with the wearing of a sword, and throughout the first and most of the second centuries, that sword would be a gladius in the same style as the ones favoured by the legions.

Auxiliary arms, frieze in Deva, Romania. (*Author's collection*)

Interestingly, as the legions become less elite, from the late second and into the third century, and the gap between the two Roman forces narrows, equipment also seems to become more uniform across both types of force. The longsword that the Roman cavalry adopted from the Gauls becomes the standard sword for both the auxiliary infantry and the legions.

Stripped of the tell-tale weapons and shields, it is more or less impossible to tell auxiliary from legionary, and a quick glance at the tombstones of the auxiliary Annaius Daverzus and the legionary Flavoleius Cordus (see comparison images earlier in this chapter) shows just how little difference there is.

The cavalry underwent an evolution in arms over the first century, quickly developing into a fairly standard kit. A shirt of chain, or occasionally scale, was complimented by a helmet completely different from the infantry models. Shields were flat, and either oval or an elongated hexagon, a design of German origin. The rider would be armed with a long sword, and with a long spear for jabbing. In some imagery, riders can be seen with short javelins, even with a quiver of the same on their horse, and even sometimes with an attendant following on behind carrying spare javelins. In Britain, though, the long spear seems to have been the standard.

Though face masks are generally associated with the *hippika gymnasia* (see below), there is evidence that metal masks were worn with helmets when on campaign. The finding of such a mask at Kalkriese, the site of the destruction of three legions and numerous auxiliary units in AD 9, suggests that it was being worn while on active campaign at the time of the Varian Disaster. We cannot say for certain whether the mask was worn by cavalry or infantry, but since later examples are uniformly cavalry items, it is tempting to apply that logic to Kalkriese.

It is interesting to note that on tombstones, especially in the first and second centuries, infantry are often portrayed in a simple stance, bedecked in just their tunic, and/or cloak, unarmoured, though sometimes with their spears. Cavalry tombstones, on the other hand, have a strong tendency to show the deceased armed and armoured, on horseback and riding down barbarians, spearing them as they disappear, cowering, beneath the horse's hooves. This, perhaps, shows us something of the difference in attitude between the *alae* and the infantry, a certain pompous pride among the horse.

We have ample evidence for the cavalry wearing breeches that reached down to below the knee, shown as much not only on Trajan's Column, but also on tombstones, and it would only be sensible for riders to wear them, even in the early days when the Roman military shuns leg coverings. The very idea of tackling a horse and a leather saddle in only a loincloth brings tears to the eyes. Over time, the infantry, both legionary and auxiliary, come to adopt leggings, both shorter '*feminalia*', then the longer '*braccae*', and eventually full, Celtic-style trousers, with leg wrappings. It is interesting to note that an examination of

Trajan's Column shows the men in *segmentata* (presumed legionaries) bare-legged beneath their tunic, but those in chain (auxiliaries) and the cavalry both wearing breeches that reach to below the knee, but also the emperor and his senior officers wearing them. While that makes sense, since the officers would be riding, just like the cavalry, it seems an impressive detail for the sculptors to have captured.

Roman infantry helmets: TL: first century 'Coolus' style ML: Principate era 'Weisenau'-type BL: Principate era 'Hagenau'-type TR: Early 'Port'-type BR: 'Intercisa'-style late Roman. (*Author's collection*)

Of course, the arms, armour and equipment of both infantry and cavalry evolved over time, and often on a converging line with those of the legions as the differences between the two unit types disappeared. Helmets tend to get heavier, bulkier and leave less and less open face showing as the second century slides into the third, and stay that way for much of the next century. By the time of Constantine, however, and the birth of the fourth century, helmets have shrunk back down to simple designs with small cheekguards and a short neck, exemplified by the variety known as the Intercisa helmet.

Auxiliary shields do not change a great deal. They broaden out over time from an oval to a circle, but otherwise there is little difference from the late first century to the fourth. The design on the shield face changes, of course, but shield imagery is a vastly complex subject too large to deal with here in more than a summary. The legions, as they change form over the third and fourth centuries, come into line with the *auxilia*, bearing the same large, circular shields and simple helmets.

Whether the *auxilia* wore *segmentata* at any point becomes moot by the fourth century, when even among the legions, *segmentata* seems to have been phased out. Now, chain and scale shirts are the norm across the board. Weaponry changes over time, though less noticeably for the *auxilia* than the legions, for though soldiers across both forces still carry a *spatha*, the main weapon in evidence is now a spear. Additional equipment is shown to us on Trajan's Column, for we see men using slings, otherwise dressed only in a tunic and cloak, and men

Auxiliary shields of varying pattern as depicted on Trajan's Column. (*Author's collection*)

with bows who are otherwise identical to the standard *auxilia* it portrays. Since Vegetius tells us: 'Recruits should be thoroughly trained in arrows', and 'recruits should be trained at loosing stones from slings',[4] we have no reason to doubt the portrayal on the column, and it is likely we are looking at standard auxiliary equipment.

With regard to the sourcing of equipment, we know that soldiers had the price of their equipment deducted from their wages, which suggests that each man was issued with a standard kit upon his entry into the unit. Given that Vegetius suggests that the army was self-sufficient, and that certainly makes sense, we have to look within the system for that source. We are aware of state-controlled *fabricae* that mass-produced military equipment, and there has been considerable debate and a great deal of research into these grand factories. Berlin papyrus 6765 seems to tell us that the *fabrica* it deals with employs *immunes*, auxiliary infantrymen, civilians and slaves, manufacturing swords, shields, bows and artillery parts. Vindolanda tablet 155 records 343 men in the *fabricae* on one particular day (25 April of an unknown year between 97 and 105), an astounding number. Given that the unit present at the time (Cohors I

4. Vegetius, *De Re Militari* 1.15–16.

Tungrorum) numbered 1,000 on paper, but probably far less in practice, we must assume that soldiers constituted only part of that figure, as per the papyrus, the rest made up of civilians and slaves. Given the mix recorded on the papyrus, and the numbers at Vindolanda, it seems likely that the bulk of the personnel were slaves, doing the actual work, with craftsmen, designers and overseers from the military, aided by civilians, in charge.

Every fort seems to have had *fabricae*, though on a different scale. The great factories (none have been conclusively proven archaeologically, though there are a number of sites deemed almost certainly *fabricae*) were large complexes with multiple rooms, based around a central courtyard. Clearly they would *need* to be immense to house in excess of 343 men hard at work. The smaller *fabricae* in forts are simple blocks, similar in scale to the barrack blocks, store houses and granaries. The clear suggestion, then, is that equipment was manufactured en masse at centralized sites, with detachments of soldiers assigned, while the *fabricae* in forts were assigned the task of upkeep, alterations and repairs to equipment.

Of course, there would have been outside sources of equipment in addition to the government-controlled factories. Claudius Terentianus, in a letter to his father (P.Mich.inv. 5391) asks for, among other things: 'a battle sword, a […], a pickaxe, a grappling iron, two of the best lances obtainable, a cloak of beaver skin, and a girdled tunic.' And in a perfect illustration of the need to supplement insufficient kit with superior versions, in another letter (P.Mich.inv. 5390), Terentianus complains that his footwear is worthless and he's having to replace them twice a month, begging his father to send him good quality leather boots. A recently discovered example can be found in the correspondence of three centurions on a papyrus from Berenice Troglodytica citing prices of goods. In it, one officer claims 'I am giving you the money, I am sending it by dromedarius', which seems to confirm that these are auxiliary, and not legionary, centurions, since auxiliary units of mounted dromedarii are well-known.

Despite such lesser quality issues, the notion that the equipment of the *auxilia* was across the board an inferior, cheaper and plainer version of legionary equipment (barring the spears and shields already discussed) does not stand up to closer examination. Throughout history (and still relevant today) soldiers have rarely simply accepted standard issue kit, often purchasing personalized or specialized items from outside sources to supplement their equipment. The idea that a unit will be manned by automata in uniform kit bears little relation to reality, and this is perfectly logical. As long as a soldier's kit does not contravene any regulations, it makes sense to obtain the best possible version of an item. In terms of quality, his life might depend upon it, after all. In terms of 'bling', who does not want a nicer item than the next man? At certain points in history this has been more challenging, but in an era when there is already huge variation over regions, further variation within units is only to be expected.

The retention of veterans' kit within units or families would also lead to the supplementing of standard issue. Good quality hand-me-downs might well be more desirable than new lower-grade issue, after all. Soldiers tended to stamp or scratch their name and unit into the more valuable items in their kit, and in some cases we find a sequence of names, suggesting that a piece of kit had passed through the hands of several owners in its time. One notable helmet, found in Alsace (a legionary helmet, but the same would hold true for the *auxilia*) bears the stamps:

Ɔ MVRRAE MARCI
Ɔ FLORI M IVL CELERIS
Ɔ RVFINI
LEG IIII
Ɔ FLAVI APER
Ɔ FLORI M IVL

The supplementing of a soldier's kit with better alternatives goes some way to explaining the wealth of decorative and high-quality Roman equipment unearthed at auxiliary sites, particularly evident among the cavalry, perhaps owing to their higher pay. The beautiful decorative helmets found at Theilenhofen and Heddernheim clearly belonged to auxiliary cavalry, the forts there of a size for an auxiliary unit, not a legion. The astounding Pamuk Mogila helmet in Bulgaria, covered in decorative scenes of gods and the Trojan war, is dated first to second century AD, and, being found in a Thracian burial mound, may well have belonged to a Thracian auxiliary cavalryman (we have already noted how the Thracians tended to return to their native land upon retirement). Closer to home, the Guisborough helmet belongs to a non-unique style, but remains a decorative example. The leather chamfrons (horse head armour) from Newstead and Chesterholm are both high quality items, and even lesser cavalry items could be extremely ornate.

And though the examples above are all cavalry, extremely decorative, high-quality infantry items have also been unearthed, such as the Bryastovets helmet from eastern Bulgaria, a second-century conical helmet covered in embossed images of divinities and, from the shape, clearly not a legionary item. Indeed, conical helmets have thus far been connected only with eastern archers, Sarmatians and, possibly, Pannonians, making this almost certainly an auxiliary item. The Kirkham shield boss in the British museum, etched with images of divinities and trophies, is circular and flat, clearly originating from an auxiliary shield. South Shields, a fort that housed both auxiliary cavalry and infantry and also a late *numerus* in its time, has yielded a very decorative helmet cheek piece, an ornate helmet handle, and a third-century sword with a brass inlay depicting Mars.

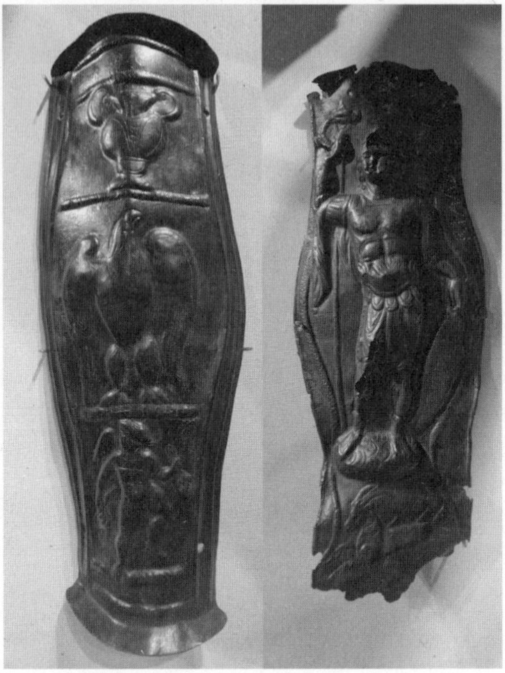

Highly decorative third century cavalry greaves at the Roman Army Museum in Greenhead. (*Author's collection*)

Decoration, then, was clearly sought and valued among the *auxilia*, and there is no reason to consider auxiliary equipment inferior to that of the legionary, which leads me to the subject of the *hippika gymnasia*. At first glance, it might seem odd that the Roman army might devote such time, effort and money to elaborate display, particularly when it involves frontline troops in provincial settings. That the glittering forces of the emperor in the capital might enjoy a little pomp and splendour in the eyes of the people feels natural; that the same might be true of a borderland cavalry unit, less so. Yet history is replete with such events, right down to our own time. Medieval tournaments were not solely the province of the high and mighty, but took place across Europe, down to a local, rural level. Even now, while the *hippika* might put us in mind of great royal events like 'trooping the colour', we only have to look to dates of military note to observe such pageantry still at work. Remembrance Day sees grand parades, and Britain is replete with annual air shows, celebrating the glory of the air force. In light of such events, it might seem less surprising that Roman cavalry regiments might stage similar events.

We do not know the purpose of the *hippika*, when and why it was carried out, though it seems likely to be tied to a religious context, as so much in Rome was. Perhaps it was connected with Tubilustrium or Armilustrium, the celebrations that marked the beginning and end of the Roman campaigning year. Or perhaps one or more of the festivals to Mars. Perhaps they were reserved for the visits of esteemed guests or high officials. Whatever the case, we are given a good description of these 'cavalry sports' events by Arrian in his *Tactica*. Putting aside the actual form of the *hippika*, which was a series of events and manoeuvres

Roman cavalry as portrayed at the Turma event in Carlisle, 2017. Note to the left the ordinary riders, to the right a draconarius, and in the centre-right, the face-plated and feathered riders ready for *hippika gymnasia*. (*Author's collection*)

designed to display the skill of the riders, in the context of equipment, Arrian is very forthcoming:

> They reach the field armoured in iron and bronze. The helmets they wear cover not only the head and cheeks, but also cover the face, leaving only holes for the eyes, and are adorned with yellow plumes that flutter in the breeze with the movement of the horses. They carry not battle shields, but lighter ones that are pleasing and decorative, appropriate to the exercise. Instead of breastplates, they wear leather tunics of a similar design with colours of yellow and blue and other colours, and leggings that fit tight (author's translation.)

In the vast majority of carved depictions of Roman cavalry, their helmets are open-faced, although they still provide more coverage than infantry headgear. The tombstone of S. Valerius Genialis in Cirencester may show him wearing a faceplate, suggested by the peak on the helmet, but since the face has worn away we cannot be sure. And yet sufficient number of cavalry helmets have been found that match Arrian's description that they can only realistically be considered helmets for the *hippika gymnasia*. In truth, given the minimal vision they allow, any manoeuvre carried out while wearing one at all would mark the man as an expert. The famous Ribchester helmet, the Crosby Garrett helmet, and the Newstead helmet are merely three of the best examples to have been unearthed in Britain, let alone examples from other regions. The Ribchester helmet predates the arrival of Sarmatians there, and seems likely to have belonged to a member of Ala

Roman cavalry helmets, from top left, clockwise: Witcham Gravel helmet, Heddernheim helmet, Straubing *hippika* helmet, Crosby Garrett *hippika* helmet. (*Author's collection*)

II Asturum. The provenance of the Crosby Garret helmet is more troublesome, but the Newstead helmet seems likely to have belonged to Ala Augusta Vocontiorum.

These helmets are generally formed of two pieces, the main helmet bowl, and a mask, often attached and hinged in the manner of a medieval visor. In first century examples, the helmet is shaped into a head with quite detailed features, but as we move into the second and then third century, the helmets and masks become increasingly more ornate, often with embossed hairstyles. They portray both male and female faces, suggesting that the *hippika* told stories in their activity, perhaps involving amazons. The inclusion of such legends and tales in gladiatorial events is well-attested, and so there is no reason why we cannot assume the *hippika* similarly involved the portrayal of ancient stories.

In terms of other *hippika* equipment, beyond the helmet, we have no clear evidence for the tunics Arrian describes, but that can hardly be a surprise, given the chances of such leatherwork surviving. They would be far rarer than, for instance, shield covers, and we have precious few of them. Plates have been found that appear to have adorned scale shirts and have been postulated as *hippika* equipment. As for Arrian's tight-fitting trousers, examples of these, reaching to below the knee, are amply attested in sculpture for general cavalry wear, and it may be that the only difference for the *hippika* was the addition of bright colours. A number of ornate greaves have been found in cavalry forts, including among the Straubing hoard of *hippika* equipment, and so, despite not being mentioned by Arrian, it seems likely they formed part of the panoply. It is hard not to appreciate the sheer pageantry and glory these events must have displayed. Given the relatively small numbers of regular cavalry in the legions,

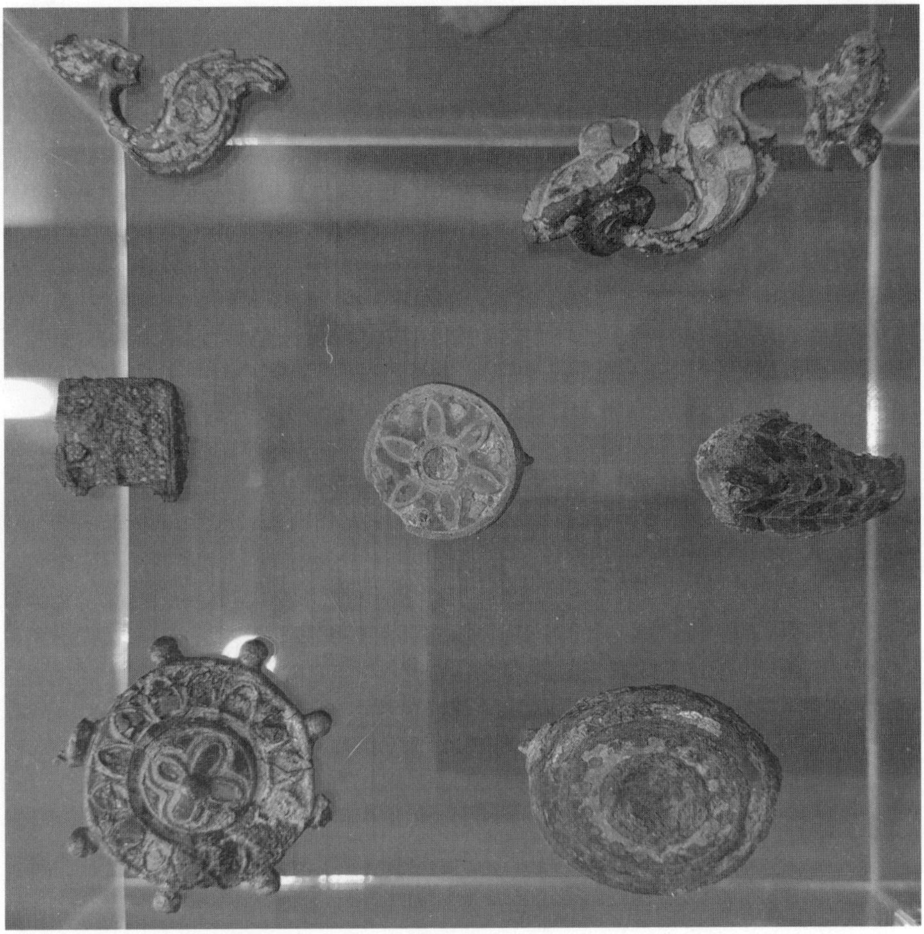

Decorative brooches in Kendal Museum. (*Author's collection*)

and the find sites of *hippika* equipment, we can comfortably say it was the *alae* that carried out the events.

In terms of personal accoutrements, the evidence from finds in forts across the Roman world shows us just how varied they could be. Torcs, an originally Gallic piece of 'bling', became such an ingrained part of Roman military life that they were even granted as awards.

It may be that during the first and second centuries, soldiers were limited in their personal gear, for Herodian tells us of Septimius Severus: 'He was the first emperor to increase their food rations, to allow them to wear gold finger rings, and to permit them to live with their wives.'[5] Forts have turned out a vast array of what are clearly fine personal possessions, and some adorned the soldiers' uniforms, as in the case of brooches.

5. Herodian, *History* 3.8.5.

While there is a certain broad similarity in equipment across the empire, for *fabricae* in various provinces would be churning out, for example, *pugio* daggers all very alike, there would also be variation, though in some ways that variation might also become a norm, that regional style spreading to become in regular use elsewhere. The sheer variety of helmet styles found across the empire, and yet the discovery of certain similar types in locations far apart makes all this abundantly clear.

So, ignoring any ethnic variations, to which we will turn next, it might be noted that in terms of equipment, barring spears and shields, there was very little difference between the legions and the *auxilia*. I am reminded of the scene in *Asterix the Gladiator* when two Roman armies meet in battle and immediately lose track of who are allies and who enemies. Clearly the only way to identify a unit in the Roman military, if armour and equipment were not an instant giveaway as they are on Trajan's Column, would be flags, standards, and the designs on shields. Sufficient pictorial evidence remains of the *vexilla* and standards, and there are a huge number of different shield designs visible on sculpture, from Trajan's Column to the Cancelleria reliefs and to other odd monuments. Moreover, the *Notitia Dignitatum* gives specific shield designs in vivid colour for a number of the units mentioned therein.

As we have seen already through plentiful evidence, not all auxiliary units were equipped the same, and some retained a solid ethnic connection. Trajan's Column shows archers in clearly non-Roman gear a number of times. They wear shirts of chain to mid-thigh, or sometimes shirts of scale, over what appears to be an ankle-length flowing robe, carry composite recurve bows, and wear conincal helmets. It is difficult on the carving to get too good a look at the sword worn by one of them, but when compared to a Syrian style sword from a Palmyran relief, they are not so different as to cause note. As such, while the column's portrayals always have to be treated with caution, we have better reason to believe these are Syrian archers than anything else. Haynes raises the possibility that the conical helmets we associate with these troops might actually be Pannonian in origin, but the Middle East and the Levant has a long tradition of such conical helmets, stretching from the Assyrians long before Rome to the Sassanians thereafter. As such, there is no reason to doubt current beliefs. It seems highly likely that Syrian archers were involved in the Dacian Wars and therefore recorded on the column, given that within just seven years Palmyrene archers are recorded in Dacia as garrisons: '*Palmyrenis sagittariis ex Syria qui sunt in Dacia superiore*' from the Căşeiu Diploma of AD 120.

Though the composite recurve bow would become the norm across the Roman empire, conical helmets are only generally found or depicted in areas with Syrian connections. In Pannonia, a notable helmet was found at Dakovo, yet in Dalmatia and in Pannonia, on both sides of Dakovo, Syrian archers could be found in the form of Cohors II Cyrrhestarum and I Augusta Ituraeorum. An example at Intercisa could have belonged to I Hemesenorum, and the image on a carving from Housesteads that stands with a recurve bow in hand, a quiver over his shoulder, and wearing a conical helmet, may well be one of Cohors I Hamiorum from nearby Carvoran. In addition to the bow and helmet, he carries a sword at his side, though the form is considerably different from those usually seen in Syrian depictions, and bears more resemblance to a later Roman eagle-hilted sword. He seems to be holding an axe in hand, although that is not certain due to the wear on the image. Most interesting is his attire. That his garb only reaches down to the knee has been taken as a clue that this soldier is not of the same ilk as the ones on Trajan's Column, in flowing robes. It might be remembered, though, that the column is a piece of propaganda, playing to stereotypes, and perhaps the long flowing robes are an affectation of the carver. Alternatively, we might equate this figure with those other archers on the column wearing scale, which seems to be only shirt length, since we cannot make out sufficient detail on the Housesteads relief to work out what precisely he is wearing.

Syrian archers on Trajan's Column. (*Author's collection*)

The upshot, then, is that Syrian archers seem to have worn chain or scale shirts (with or without the long robe beneath), carried a sword at their side, a quiver on their back, used a recurve bow, and wore a conical helmet. Of all the auxiliary forces, they are perhaps the easiest to identify from their equipment. Interestingly less than 5 miles from Carvoran, one of the known bases of I Hamiorum, a leather quiver was retrieved, now in the Hancock museum in Newcastle. While the quiver could have belonged to an auxiliary soldier stationed at Birdoswald, it is tempting to see it as the presence of a Syrian archer from the next fort along on the wall.

We have already met the Sarmatians in our discussions: a thoroughly non-Roman people born of the steppe and brought en masse across the empire to an unfamiliar land. Given the nature of their 'recruitment' in AD 175, it is perfectly logical that the Sarmatians would have retained more of their individual ethnicity than any of the other peoples formed into auxiliary units in Britain. We cannot be certain how the units were equipped as they were assimilated, but given how utterly different they were to any other horsemen within the empire, it seems quite unlikely that they had their pay docked and were supplied with standardized Roman equipment from the *fabricae*. Almost certainly they wore their own traditional armour, used their own weapons, and continued to manufacture them even as they now served Rome.

On Trajan's Column, the Sarmatians are shown wearing conical helmets with reinforcing ribs, and full body-suits of scale, including long sleeves and trousers, their horses similarly attired in all-covering scale. Their weapons are missing, given that they would have been bronze lances that have since disappeared, but the position of their arms suggests a weapon of some length. One rider is shown with a small composite bow, loosing back over the rear of his horse in the manner of the Parthians. Behind one group, an unarmoured Dacian carries their standard, the infamous *draco*. As always with the column, we need to approach the image with care. The artisans certainly had never laid eyes upon a Sarmatian, and would be working from hearsay descriptions. The simple logistics of riding a scale-clad horse in scale-clad trousers and, apparently, without a saddle, are eye-watering and clearly fictional, as is the idea that horses could be fitted with skin-tight legs of scale armour. As such, it seems likely the images are largely the product of an artisan's imagination, based upon a loose description.

However, the gravestone of what appears to be a Sarmatian rider, found at Chester, sadly with no surviving inscription, echoes the column in some ways. The *draco* standard is there, and the helmet shown is much the same shape as those on the column. Due to the condition of the surviving stonework, we cannot say for certain what the armour consists of, but the carving bears a marked resemblance to another image of a Sarmatian rider from the city of Tanais on the Russian shore of the Sea of Azov. That rider also wears the conical helmet, has a cloak like the Chester horseman, a long *contus* lance, and is clad in a scale shirt that has half-length sleeves and reaches down to just above the knee. Similarly, we cannot make out much detail of the horse on the Chester carving, and the Tanais one is not clear, but the animal's head and neck may suggest armour of some sort. A wooden figurine also showing a Sarmatian again has scale armour, and so we can only conclude that scale was the standard armour of their warriors, though perhaps not as all-encompassing as the imaginative sculptors of the column would suggest. An interesting insight into this armour comes from Pausanias:

> Among the votive offerings there is a Sauromatic breast plate. [...] Their breastplates they make in the following fashion. Each man keeps many mares, since the land is not divided into private allotments, nor does it bear any thing except wild trees, as the people are nomads. These mares they not only use for war, but also sacrifice them to the local gods and eat them for food. Their hoofs they collect, clean, split, and make from them as it were python scales. Whoever has never seen a python must at least have seen a pinecone still green. He will not be mistaken if he likens the product from the hoof to the segments that are seen on the pinecone. These pieces they bore and stitch together with the sinews of horses and oxen, and then use them as breastplates that are as handsome and strong as those of the Greeks. For they can withstand blows of missiles and those struck in close combat.[6]

It has been suggested that the very name 'Sarmatian' derives from the Greek word for lizard (*sauros*), which might describe both the *draco* standards and the scaled appearance of the warriors.

The lance shown in depictions of Sarmatians, known as the *kontos* (Latinized to *contus*), was an increasingly widespread weapon among the Roman *auxilia* from the second century onwards, but it is worth noting that they do not appear until the time of Trajan, who, coincidentally, battled and overcame Sarmatians in the Dacian Wars. It is generally accepted that the *contus* was a Sarmatian weapon adopted by the Romans, and so we can see it in Britain as still a Sarmatian cultural weapon. The *draco* standard, which, again, became an integral symbol of Roman cavalry in the third and fourth centuries, seems to have been a Sarmatian design, adopted over time by others.

6. Pausanias, *Description of Greece* 1.21.5–6.

That the Chester Sarmatian bears the ubiquitous conical helmet and carries a *draco* suggests that they continued to cling to their cultural symbols and equipment even as a Roman unit posted halfway around the world. A now-lost tombstone of a decurion from the Ala Sarmatarum in Ribchester is recorded also as showing a rider with a two-handed lance (*RIB* 595).

The sometime allies of the Sarmatians, the Dacians, are also represented in Britain's *auxilia*, as we have seen, in the form of Cohors I Dacorum at Birdoswald. It is interesting to note that despite several excavations at Birdoswald and the longevity of the cohort's presence there, no item has been unearthed that shows a distinctly Dacian origin. It would appear that I Dacorum were armed and equipped in standardized Roman equipment from the *fabricae*. One tantalizing hint of their clinging to native symbols can be found in the form of the two inscriptions from Birdoswald mentioned earlier, both solidly Roman-style inscriptions in Latin, yet both decorated with images of the *falx*. No evidence of their use has appeared at the site, but their appearance on images is interesting and shows the value of the native armament as a symbol, in much the same way as British army regimental insignia persist in the use of swords in their design despite the weapon having long fallen out of use.

The last group to examine in terms of ethnic equipment is the Mauri. Though the only recorded presence of the Moors in Britain comes from the Numerus Maurorum Aurelianorum, mentioned in the *Notitia* as based at Burgh-by-Sands, we cannot deny the likelihood of their presence in Britain at least since the Severan campaigns, if not before. Finding evidence of their cultural equipment is problematic at best. In fact, the absence of evidence is, on this occasion, almost evidence in itself. The Mauri are well-attested in imagery on Trajan's Column.

Mounted on small ponies, the Mauri are shown unarmoured, riding only in loose-fitting tunics, bare-legged and bare-headed. Their hair is distinctively fashioned in almost dreadlock fashion. Of course, their thoroughly un-Roman ethnicity is probably deliberately heightened by the artist as part of the propaganda. Their armament, though, matches what we are told by ancient writers. They carry small, round shields, and though their weapons have disappeared from the column over the centuries, from the position of their hands, the spears they carry cannot be long, and would fit the description of the javelins we hear of in sources. Their light kit, almost the polar opposite of the heavy Sarmatians, perfectly suits their use as light, skirmishing cavalry.

Leaving aside arms and armour, the question of uniform is a subject of study in itself. Certain units, of course, seem to have kept their own ethnic variations – the Mauri and their light, billowing tunics, the Sarmatians' cloaks, the robes of the Syrian archers. But the vast majority of images show the Roman army in a very standardized manner.

We have already touched on the matter of legwear. In the early days of the Principate, the Roman infantry is shown uniformly bare-legged beneath their tunics. Cavalry are shown in tight-fitting breeches that reach below the knee. What becomes apparent is that such breeches are gradually adopted by the infantry, probably earlier in the northern provinces, for the dual reason of being constantly exposed to such clothing among the native peoples, and because of the weather. It is easy to picture the Roman army in a standard kit, but it should always be borne in mind that only a fool would not adapt to suit local conditions, and that the natives of a region have adopted certain clothing with good reason. As such, we find that as time wears on towards the end of the second century, the wearing of some sort of breeches beneath the tunic becomes common, though not total. By the time of the Dominate, in the late third century, trousers are more common. Finds of such things, of course, are rare, though examples excavated particularly in Germany and the Netherlands give us some ideas of their form, beyond the images on imperial monuments.

Knee-length breeches of the late first or second century in Dusseldorf are of undyed wool, and an example from Valkenburg is of leather. The Damendorf trousers, dated to the second to fourth centuries, are woollen and full, ankle-length, again in a natural hue. Indeed, there is ample evidence to show that by the later fourth century, full trousers, or thick hose at least, were the standard. The diptych of the great general Stilicho shows him wearing patterned leggings with feet attached, within the shoes. The fourth century mosaics of Piazza Armerina show a wide variety of figures in full trousers, but perhaps best evidence of all is a wall painting from a tomb in Silistra, Bulgaria, dated to AD 375, which shows a slave carrying his master's trousers (again with built-in feet) over his shoulder.

Though we have little evidence in Britain, given that the majority of the Roman military on the island served either in Wales or in the north, and given the weather conditions for much of the year, it is likely that legwear was adopted both sooner and on a larger scale than in much of the empire.

Tunics are less prone to change over time. There are minor variations, but from the early Principate to the Dominate there is little visible difference, beginning as essentially a sleeveless woollen bag that is given shape with a belt, worn to above the knee, a fashion that denoted a military man over a civilian. The form seems to have altered gradually to allow a little more shape and variation, though never far from the standard. It is only by the third century that things begin to change. Long sleeves begin to appear and by the fourth century tunics have become much more ornate in decoration, often with complex rondels or other designs on sleeves and above the hem. The colour of the military tunic has been hotly debated over the decades. A number of the portraits on the Fayuum mummies portray soldiers, according to Graham Sumner, who has studied the

Part of the Nile Mosaic from Palestrina, showing soldiers in off-white tunics. (*Author's collection*)

subject in detail. On the whole, their tunics are of a basic off-white, but appear to host coloured decoration, including a zig-zag, triangular pattern around the neckline. We know that the wearing of narrow or broad stripes on a tunic denoted a man's rank, and therefore plain white would make such rank insignia stand out clearly. On such portraits, as well as the colour of tunic decoration, Sumner also notes a variety in the colour of the cloaks worn over the top (red, blue, purple, ochre), and suggests that a combination of colour and design upon the tunic and cloak could allow for a wide variety of ranks being portrayed.

Servius Ennius Fuscus wearing a *paenula*, with his wife, on a tombstone in Split Archaeological Museum. (*Author's collection*)

Of course, we cannot say whether these figures are legionaries or auxiliaries, and since some of the Fayuum portraits stretch into the fourth century, the question becomes not one of legion versus *auxilia*, but of field army versus *limitanei*. With precious little surviving military clothing (and the bulk of that being from the eastern empire) we can say little with certainty about the everyday tunic colour and décor of the *auxilia* in Britain.

Cloaks are more readily identifiable. In the Principate, they come either in the form of the *sagum*, a square that is gathered and pinned at the shoulder to form a cloak, or the *paenula*, a semi-circular cloak of more familiar design that could be fastened at the throat. Cloaks would be of wool, or of felted wool, both of which would provide natural waterproofing from the inherent lanolin, although an added layer of waterproofing would be likely. Some cloaks may have been fashioned of leather, especially given the British weather. The *paenula* is last attested in imagery in Britain by the mid-second century, with the *sagum* becoming the standard, along with the eventual introduction of the Gallic *caracalla*.

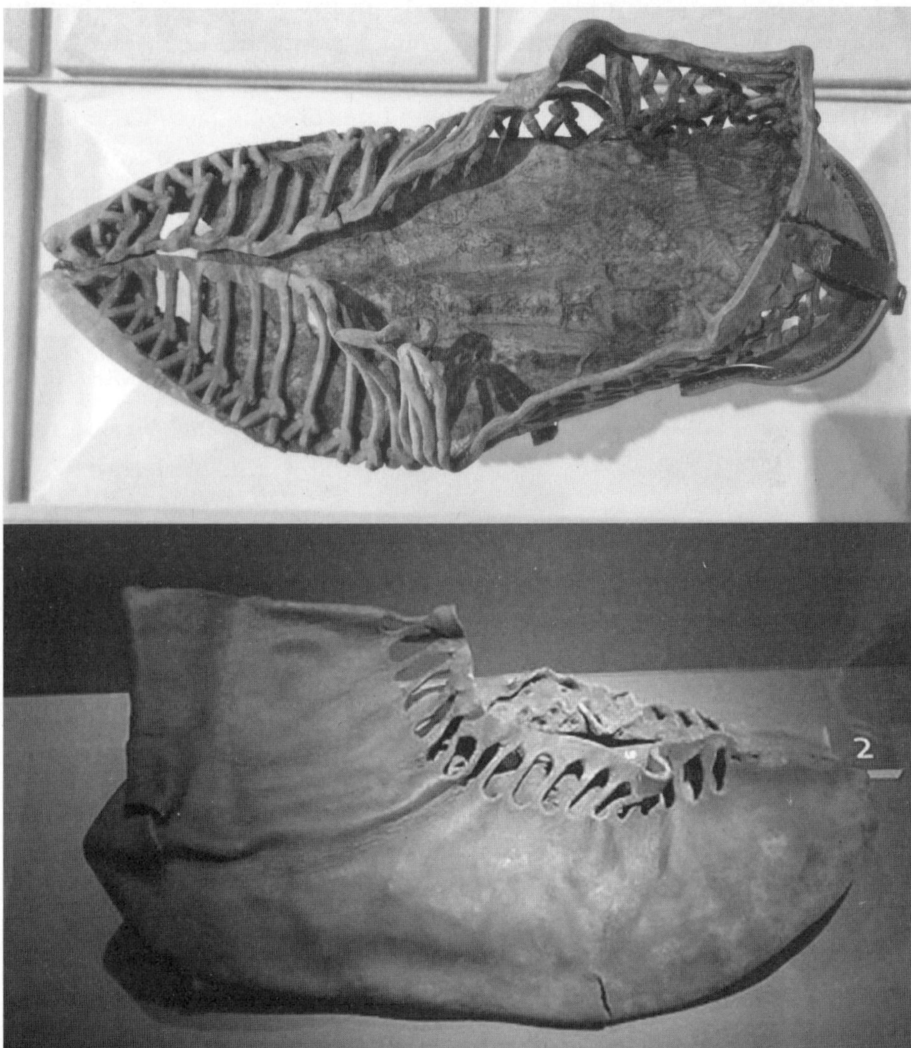

Roman caliga and enclosed boot, Hancock and Vindolanda museums. (*Author's collection*)

A view was long-held that the armies of Rome all wore the famous *caliga*e sandal-boots, but the impracticality of such footwear in a northern British winter is clear, and actual evidence of their use in Britain is sparse. Their use in general seems to peter out in the same period as the *paenula*, to be replaced by a more practical enclosed leather boot with hob-nailed sole, a huge variety of which have been unearthed in recent years, particularly at Vindolanda, highlighting their commonality. Boots were worn with woollen socks for warmth and comfort, as we are reminded by Vindolanda tablet 346: 'I have sent you [...] pairs of socks from Sattua, two pairs of sandals and two pairs of underpants.'

As previously noted, however, while there would be for most recruits a standard uniform provided by the army at the expense of pay deductions, it is hard to imagine any soldier not saving up a few coins a month and then splurging at the market and buying something special to adorn his kit. An imported pair of Italian boots, a thick, heavy Gallic cloak, a tunic of Egyptian linen, perhaps. And while it might be hard to imagine a wealth of foreign goods reaching some rainy corners of northern Britain, it is worth remembering that the amber used in beads found across Britain will have come from the Baltic via the great trade routes down to northern Italy and then back out to the provinces from there. Roman trade was expansive and impressive, and if a soldier fancied a little something oriental, be sure he could get it.

Once again, the only standard thing about the Roman military is how flexible it could be.

Chapter 5

Praesidio

(Garrisoning the *auxilia*)

Auxiliary forts in Britain are the subject of an entire branch of Roman archaeology, and constitute both the best visible remains of Roman presence on the island, and the best source of archaeological knowledge. From the early excavations of men like John Clayton to the ongoing project at Vindolanda, the evidence drawn from the ground adds to and changes our knowledge of Roman military history at a surprising pace. Study of the Roman forts of Britain and their history is tied inextricably to the auxiliary units who manned them.

The first thing to note about the many defensive fortifications discovered across Britain is their purpose and nature. It is tempting, as always, to see Roman culture as consisting of clearly delineated systems, but, as should now be apparent, what appears to be a rigid system can often prove to be flexible. Still, we can categorize Roman fortifications into several forms, and concentrate only on the ones that concern the *auxilia*.

The largest such installations in Britain, and sites that still draw a huge number of visitors, are the legionary fortresses. The purpose of these should be clear: as main bases for the legions in Britain. From the late first century, those fortresses number just three: Caerleon, Chester and York. Evidence for earlier fortresses abounds, sites that were manned for a while before moving on: Exeter, Inchtuthil and Usk are examples. These sites are all immense, as one would expect when housing over 5,000 men, cavalry detachment and ancillary staff. They are also out of our purview. There may have been occasions when auxiliary forces were temporarily billeted in these places, but they were never garrisoned there.

Another type of fortification is the temporary camp. These remain, where there is any evidence at all, as simple earthen banks, usually in the familiar 'playing-card' shape. There are a number of reasons for such works: as marching camps, constructed hurriedly to provide over-night protection for a force on the move; as practice camps, built simply to practise construction or to use for other drills and tutorage; or as temporary accommodation during construction

Ramparts of turf and timber, reconstruction at Park in the Past. (*Author's collection*)

of other sites, and undoubtedly for many other reasons. They consisted solely of protective ramparts and often ditches, the soldiers living in tents within the banks. The larger examples of these sites are often seen as 'vexillation fortresses', temporary camps for large armies on campaign or gathering to prepare for such. The size of temporary camps can vary hugely, from small examples to house a cohort of 500 men to vast hillside fortresses to hold more than one legion and their auxiliary support. There is a troublesome bleed area between two categories here, for there are also permanent forts that were only ever constructed of earth and timber, and identifying whether an unexcavated site was a temporary camp or a turf-and-timber fort can be difficult.

There may be a reason for this. Throughout the first century there was no true notion of an imperial border, especially in Britain. It is only with Trajan's fortification of the Stanegate, and then Hadrian's expansion of this into his wall system, that Britain becomes a permanent border province. Until this point, it was never a question of whether the entire island would be conquered, just when. As such, fortifications were *all* temporary, for the military would move ever northward until Britain was Roman, and once it was settled, most of the military would be moved to other provinces, leaving only a caretaker garrison. Thus, until Trajan's time, all camps were more or less temporary, and built in wood and earth, the exception to this being the legionary fortresses. It might be remembered that while in the modern world it is extremely costly and troublesome to move military units from one base to another, in the Roman world those units are self-supporting, able to build their own fort, supply their own raw materials and even shape them into usable goods, and so the moving of a unit from one site to another is not quite the massive event it might sound.

Even into the second century, when it becomes apparent that the island may permanently house a border and therefore need garrisons on hand, those forts are still largely constructed of timber. The clear exception to this is the forts of the Stanegate and the walls of Hadrian and Antoninus as they are constructed. Other forts, either beyond the walls or in the hinterlands, largely remain of timber construction for some time, and those that are finally rebuilt in stone are often attributed to the Severan era. Thus there is a sequence: temporary camps stop being such when a unit is assigned as a garrison, but remain in the same form for some time before any rebuild. Not all forts were ever rebuilt in stone and some remained of turf and timber.

Of the forts that were constructed in stone in due course, not all housed stone buildings inside. In some cases only the walls and the central buildings were of stone, the barracks and stores and the majority of internal structures remaining wooden. The exception to this is the bath house, which we will examine shortly. It is worth noting also that despite the general uniformity of Roman forts, not all fit into the standard mould. Forts were occasionally of entirely different shapes, in order to fit the terrain.

It is these permanent forts, smaller than the legionary fortresses, but used as a garrison rather than a temporary site, that we can associate with the *auxilia*,

Fort shapes (from top left, clockwise): Vindolanda, Whitley Castle, Lunt, Bewcastle. (*Author's collection*)

as well as a number of smaller sites (fortlets) that start to crop up from the second century.

Forts were commonly constructed around 10–15 miles apart, an easy half-day march. There are exceptions, of course. Some forts were forced closer together by terrain, or logistical needs, to guard fords or bridges, at important crossroads for example, and the forts on Hadrian's Wall are generally less than 8 modern miles apart. Forts would be linked by good military roads built by the legions and (as we can now surmise) also the *auxilia*. We are fortunate to have a surviving source on Roman forts in the form of Hyginus's *Liber de Munitionibus Castrorum*, in which he gives us the criteria for the siting and construction of Roman forts. Of course, as always, Hyginus concentrates principally on legion-based armies, but the text gives us clear details for forts in general. Rome bends rules, and Hyginus is only giving us a guide for the perfect fort. We do not have to look far to see his advice overlooked: 'No mountain should loom over the camp' hardly applies, for instance, to the fort of Hardknott.[1] The siting for installations, in the end, came down to the commander in the field, assisted by engineers and architects, locations in unfamiliar territory searched out by scouts. Tacitus tells us of his father-in-law, Agricola: 'It was noted by experienced officers that no general had ever shown more judgment in choosing suitable positions.'[2] Of course, even Agricola was not infallible. There is ample evidence across Britain for forts having been re-sited at some stage, an excellent example being Dalswinton, a site seemingly selected by Agricola by the River Nith, which soon had to be moved to higher ground due to flooding.

The siting of forts varies according to terrain and need. Some are on prominences (such as Fendoch, Bewcastle, Whitley Castle), others on open flat ground (Watercrook, Ambleside, Piercebridge). Forts are almost always close to a water source, and often at the confluence of two. They are commonly to be found where a major Roman road crosses a sizeable river.

When the fort site was chosen, the defences were raised. We shall use as our example the more standard Roman fort, of the familiar playing-card shape, but bear in mind that the shape and form could vary, and this is simply the most common format. Ramparts were raised, usually by digging a defensive ditch and using the spoil to raise a mound within it, often strengthened with timber or stone bases, leaving a short flat space, a berm, between ditch and rampart. The rampart would be given the steepest possible exterior face, and was then fortified. There is still considerable discussion over precisely how the defences

1. Hyginus, *Liber de Munitionibus Castrorum* 57.
2. Tacitus: *Agricola* 22.

Timber Dacian ramparts on Trajan's Column. (*Author's collection*)

of a timber fortress were formed, varying from palisades of heavy logs to woven wicker fences, but whatever the precise form, a wooden fortification of some form sat atop the earth bank. Another sign of the inaccuracies of Trajan's Column is that the legions are shown throughout the frieze constructing stone forts, which is extremely impractical and unlikely on campaign in a war. Indeed, there is only one scene on the column that appears to show timber fortification, but the defences, clearly a palisade, are Dacian, being fired by Romans.

Each of the four sides of the fort housed a gate. Hyginus suggests that the fort's main gate, the *porta praetoria*, should face the enemy, though in practice it would often be sited to face either the principle access route or a bridge over a river. The rear gate, the *porta decumana*, would be directly opposite on the far side of the fort. The other two gates, the *porta principalis sinistra* and *porta principalis dextra* would lie to left and right, with the main gate ahead. On Hadrian's Wall we find a variation, since some of the forts lie astride the wall rather than behind it, and so sometimes there are two gates in each long axis, one north of the wall, one south (examples at Benwell, Rudchester, Halton Chesters, Chesters and Burgh-by-Sands). Some forts' layouts remain unknown, while Birdoswald sits south of the wall but has the extra gates dating from its position astride the original turf wall, and Wallsend has an odd collection of gates, given its position at the end of the system. Over time, the wide turf

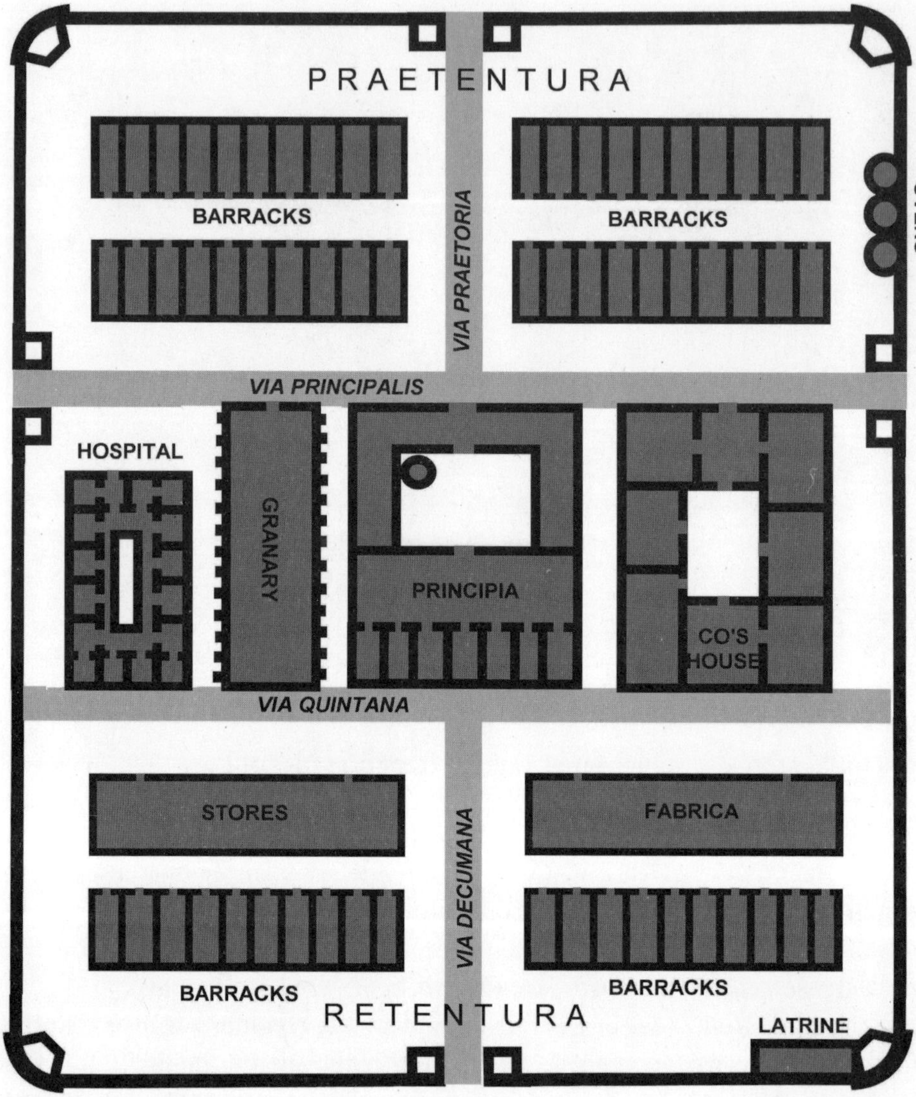

Plan of a typical Roman fort. (*Author's own work*)

mound of the rampart, even when externally faced with stone, would come to house ovens and latrines.

The internal organization of the fort followed a fairly standard form (though as always with variation where required). The two side gates were connected by a road, the *via principalis*, that cut across the fort around a third of the way from the main gate to the rear. A second road ran across some two thirds of the way down, connecting the secondary side gates when they existed. Between these two crosswise roads lay the main administrative buildings of the fort, in some

Reconstructed *principia* at Saalburg, looking at the chapel of the standards. (*Author's collection*)

cases the only internal buildings of stone. At the centre stood the *principia*, the headquarters building, looking down the *via praetoria* that led to the main gate. The precise form of the *principia* varies from fort to fort, but the principle features consist of a courtyard, often with a colonnaded walkway around the edge, and sometimes rooms. In the courtyard was often a well, and perhaps imperial statuary and altars to the unit's gods. A hall, the *basilica*, stood often on the far side of the courtyard, from which led off a series of small rooms. The rooms in the *principia* would house the offices of the commander and his senior men, the armoury, and the most important room in the fort: the *aedes*. This room, the 'chapel of the standards' would house all the standards of the unit, the fort's most prized and valued items. Moreover, it was beneath this room that we usually find a subterranean strongroom, in which the unit's pay chests could be stored. Some *principia* had a large hall across the front of the building, often associated with cavalry forts.

The *principia* is flanked by buildings, to one side the commanding officer's residence, usually a spacious courtyard house, sometimes even with its own small *balneum*, or bath suite. The commander's house at Chesters contains an impressive example, and Chesterholm, with a sequence of forts on the same site, boasts the remains of two separate such houses, each with their own baths. On the far side of the *principia* stood, usually, a pair of granaries, storing the fort's foodstuffs. Sometimes there was but a single granary. Sometimes an extra building might be found here, of unknown designation, perhaps a workshop or stores. Occasionally a *valetudinarium*, or hospital, might be found here, as at Wallsend. The rest of the fort, the area closer to the main gate being the

Sallburg fort, Germany: Reconstructed granaries, barracks and *fabrica*. (*Author's collection*)

praetentura, that to the rear the *retentura*, contained barrack blocks in ordered rows, with stores and stabling as required. The precise form of the barrack blocks varies according to period, unit type and probably the personal preferences of the designers. The most commonly recognized form throughout the first and second centuries is a line of ten rooms, each designed to house an eight-man contubernium, with a larger room at the end for the centurion, though this is subject to immense variation.

Hospitals are not a feature of every fort, but have been identified at a number of sites, including Fendoch, Wallsend and Housesteads, with a record mentioning one at Chesterholm, too. These structures were designed as a series of individual rooms around a square, connected by a corridor, usually with a garden or courtyard in the centre, its own water supply, latrine and drainage. They have been identified as hospitals not only by their form but also by the medical equipment found in them. Given their sporadic appearance in forts, it may be that each *valetudinarium* was expected to handle not only the garrison of that fort but also a catchment area of other local garrisons.

One concern for every fort would be water. In addition to a well, there would need to be sufficient running water for an entire cohort, including a hospital if present, flushing latrines, and feeding the bath house that was a feature of every fort. Many forts were sited in positions where they could make use of springs and streams, diverting the flow to supply running water for the fort. In such cases it is not uncommon to find springs becoming cult sites, such as Coventina's Well at Carrawburgh. In other situations, aqueduct channels were carefully planned to bring water to the fort from the nearest viable source. Carvoran and Lanchester are examples of forts with extensive aqueduct systems, and we know of others from epigraphic evidence, such as the aqueduct of Caernarvon mentioned on *RIB* 430. With forts that sit upon ridges and have no source of water from a higher position, other solutions were required. The planners of Housesteads placed its baths some way from the fort, by the Knag Burn, while six known stone water reservoirs have been found so far, suggesting that collected rainfall provided most of the fort's fresh water, and overflow could be used to flush the latrines that lay in the lowest corner of the fort. Hardknott, despite its lofty position, could count on streams flowing down from the peak. How forts like Fendoch or Bainbridge (both on high platforms) managed to acquire fresh water remains to be found, though clearly they did. Perhaps, like Housesteads, they relied upon collected rainwater.

One building that was constructed in stone from the beginning, regardless of the rest of the fort, was the bath house, for obvious reasons, given the presence of underfloor heating and furnaces. I will not venture into a diatribe on the nature, form and use of Roman baths here, but suffice it to say that bathing was

Water tank at Housesteads. (*Author's collection*)

an essential part of Roman culture in general and would, of course, be of critical importance for health in the army. Each fort had its own bath house, varying in size and complexity, and often the baths are found to have been rebuilt and enlarged from time to time. Hardknott's simple baths consist of a single range of four chambers, from changing room, to cold, warm and hot rooms, with a circular steam room, or *laconium*, separate from the main structure, while Chesters boasts a large changing room with cubby holes, a double steam room, latrines, and special apsidal baths.

In addition to these standard structures (and occasional extra hospitals), forts could boast a number of other buildings, sometimes identifiable, sometimes not. Temples abound, particularly good examples being visible at Benwell and Carrawburgh, and it appears that some forts even boasted a small amphitheatre outside their walls. (Tomen-y-Mur remains impressive even now, while those of Newstead and Inveresk remain buried.) The Lunt fort (one of the most unusually shaped in Britain and which was never rebuilt in stone) hosts a circular wooden structure within the ramparts that has been identified as a *gyrus*, a horse training circle. A number of forts have *mansios* (imperial waystations) outside their walls. Parade grounds have been located outside several forts and may have been a standard feature. The one at Hardknott is a particularly interesting example, for it has been levelled out of the hillside, with a tribunal mound at one corner for the officers to mount for observation. Seemingly quite speedily,

The spacious changing room (apodyterium) of the fort bathhouse at Chesters. (*Author's collection*)

civilian settlements grew up outside the forts, hosting taverns, shops and traders, brothels, workshops and more, taking advantage of a captive audience to relieve the garrison of their wages. Over time, these settlements became more and more connected with the forts, with soldiers having unofficial wives in the extra-mural buildings, and in a number of cases, when a fort was abandoned, the settlement spread across its site to become a fully-fledged town in its place.

When fort walls were rebuilt in stone, sometimes during the Hadrianic and Antonine periods, but often in the Severan era, gates almost always had twin towers adjoining them, though in a few cases single towers have been identified. Many forts have also shown evidence of early timber towers around the circuit, and when rebuilt in stone, towers become standard. For the majority of the period, towers are internal, the outer face flush with the wall, though in the later period, they begin to project beyond the walls, giving good enfilade fire possibilities. This may show a gradual shift in military thinking from a fort being a garrison from which forces would issue for warfare, to a more defensive site. We can often tell the unit that was involved in the rebuild from inscriptions left to mark the work. In some cases, only the walls and the principal central buildings were given stone replacements, with barracks, stores, workshops and the like continuing to be of timber. (Hardknott and Ambleside are prime excavated examples.)

From the Severan era onward we begin to see more and more variation in the fort's internal organization. South Shields shows a rebuild on a massive and very specific scale, a smaller fort being extended and many of the barrack blocks removed to make way for row upon row of granaries, believed to form part of Severus's grand plan for the invasion of Scotland. The later iteration of the fort at Binchester seems to buck the general trend in being split in two by Dere

Hardknott and Ambleside forts from the air. (*Images: Google Earth*)

Street, with a series of barracks filling one half on the long axis, the other side showing an unusual arrangement, including several iterations of commanders' houses. At Caernarvon, while the commander's house does not seem to have had its own bath house, it was a large example, but in the third century a large courtyard house was built in the southern corner of the fort with its own baths, presumably over the remains of barrack blocks, and even this gave way to a new bath house in the fourth century.

As always, then, variation and flexibility are the hallmark of the Roman military. In addition to the auxiliary forts, Britain plays host to a large number of small fortlets (also prevalent in Germany). Excavation of some of these have given us considerable information of their layouts, and in some cases they have been found occasionally to have supplanted full-sized forts (Castleshaw and Chew Green) or even to have been replaced by a larger fort (Ravenglass). Since they invariably date from the Trajanic era onwards, it is tempting to see such fortlets as the home of the *numeri* that began to be recruited from this time on, although it is not impossible that they played host to vexillations of cohorts or *alae*, or even legions, based at other sites as small caretaker garrisons.

There are many changes over the later third and fourth centuries. Beginning with the Severan decision to allow soldiers to marry, we see a gradual shift of military thinking that seems to follow a sequence. Over time, gradually, as the size of garrisons reduces, forts are reorganized, perhaps so that a soldier's family can be moved inside the walls with him. This might be the cause for a general move away from rows of communal barrack rooms to lines of individual chalet-style buildings. Certainly the division between the military and the civilians seems to close, with legionary compounds appearing in the heart of the Roman town at Corbridge in the third century. In the fourth century we also see a sporadic decline in the civil settlements outside forts, with some disappearing altogether. It is tempting to suggest that this ties in with the civilians being brought inside the fort for better protection.

As Britain approached the end of Roman control, increasingly the focus fell away from the fortification of the north and the danger of the Picts (despite repeated disasters such as the 'Great Conspiracy' in which tribes flooded into the poorly protected island). At this time, the principal danger Rome faced in Britain was from overseas, either the raiding Saxons and their ilk from northern Germany and Denmark, or from the Irish tribes. As such, new systems of defence were created. The basis of this was already being put in place in the early third century with sites such as Reculver, but over the third century and into the fourth, the system slowly grew to become a border of its own to match the walls of Hadrian and Antoninus. Whether there was an intention from the third century onwards to create a full system of coastal defence, or whether it was a periodic growth to protect danger spots that was combined in the late fourth century into what we now see, remains to be seen. There is still a strong tendency to see this system as separate parts:

- The Saxon shore
- The Yorkshire coastal signal stations
- The coastal defences of Wales

However, when one looks at the fortifications that were in place by the close of Roman rule, it is an impressive feat to match the greatest of border fortifications historically – to ring in an island nation with fortifications and warning stations.

Beginning with Hadrian's Wall, which was still occupied in 410, that system seems to have extended down the Cumbrian coast to at least Ravenglass. A late Roman shore fort exists at Lancaster and onward lies Caer Gybi at Holyhead, with its signal station on the mountain above. It may be that these three locations were deemed sufficient to protect the wide expanse between Wales and the Lake District, although it is perhaps more likely that other sites that link these together remain to be identified. There is a similar problem moving south from Holyhead. Late Roman installations have been found at both Caernarvon and Cardiff, but there is an enormous amount of coastline between the two sites, which faces Ireland, and so it seems likely there were further sites there too. The coastal stations identified in Cornwall seem to belong to the period of the invasion, rather than late Roman Britain, and therefore are unlikely to be part of any late system. Across the south coast, though, Britain was sufficiently

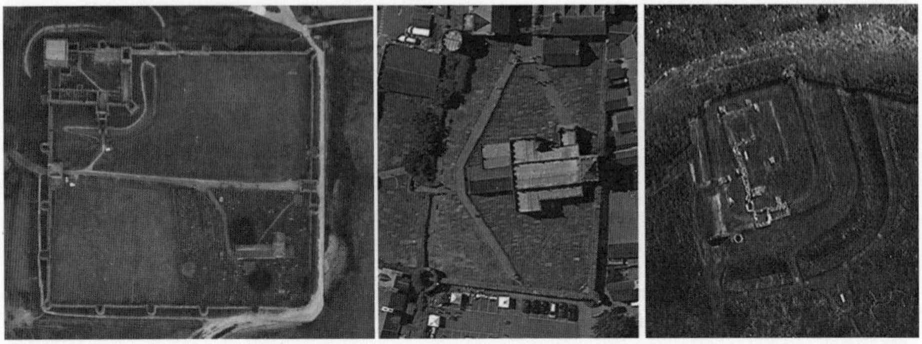

Installations of the Late Roman coast: Portchester Saxon Shore fort, Holyhead fortlet and Scarborough signal station, from the air. (*Images: Google Earth*)

urbanized and fortified, and when combined with the fact that this stretch of coast faced only Northern France, which was equally fully Romanized, it is likely that no new defences were considered necessary here. The Saxon Shore then picks up at Portchester, and creates a solid coastal system all the way to the Wash, and even mid-Lincolnshire, based on the assumption that Leland's identification of a lost Roman site eroded out to sea at Skegness is accurate. There is the distinct possibility that the system went further and joined with the signal stations of Yorkshire, and that coastal erosion has simply removed all traces of them, as it now threatens to do with the ones that remain. Sites all the way from Filey to the Humber have unearthed late Roman artifacts that might suggest the presence of now-lost signal stations. If we were to suggest that a similar fate has fallen upon other locations north of the Tees, then that system would link up with the end of Hadrian's Wall, effectively sealing the island to outside offensives. A series of stones on the coast at Roker, near the mouth of the Wear, may suggest another coastal station from the period, now lost to the sea, supporting that notion. Such a grand system of coastal defences was not to be seen again until the Tudor era.

This system fits the later Roman army, of course. We have already seen how successive removals of manpower across the Channel would have reduced the army in Britain, perhaps to a critical extent, towards the end of the fourth century. We can deduce from the reduction of legions to mere coastal garrisons, from the proliferation of small units such as *numeri* and *cunei*, from what is understood of the changes in the entire form of the Roman military, that in the fourth century, the Roman forces in Britain consisted of many small units rather than fewer large ones. The late Roman shore forts, fortlets and signal stations would be appropriate garrisons for such units. This is not the place for a full investigation of late Roman coastal defences, though I would point the interested reader towards *The Roman Shore Forts* by Andrew Pearson.

Chapter 6

Last Thoughts

Before we look at the histories of, and evidence for, the auxiliary units themselves, a few points could do with reiterating or clearing up. I am aware that I have made the point more than once already, but with an eye to the unit histories that follow, it should be borne in mind that the Roman systems are flexible to an impressive degree. Tribunes could serve as prefects; any level of officer could be assigned a command under the title of *praepositus*; forts were always playing-card shaped and with the same layout, except when they were not; a fort held a single cohort or *ala*, except when it didn't, that unit possibly being split between more than one garrison site. There are rules to everything, but not everything follows those rules.

Secondly, building on that important point, it is worth remembering that much of what we can deduce about the Roman military in Britain is drawn from the written word, be it in ancient texts, on wooden tablets that had been disposed of, or on altars and inscriptions. While we are forced to take these at face value, we must always be aware of the possibility of error or misinterpretation. When studying Agricola, I became acutely aware of how a different translation of Latin could entirely change the meaning of a sentence – '*firmabatur*' is one such word that plagued me – and so while I have taken the best meaning I can find from any ancient text, there may always be alternate views. A similar problem occurs with written sources such as the Vindolanda tablets. With them, we are entirely at the mercy of the writer, and if they made a mistake, then all our research based on it is also mistaken. *The Antonine Itinerary* is plagued with problems that may have arisen from some copyist over the centuries mistaking an X for a V or the like. We have an even more fundamental problem with inscriptions, and in particular with altars. The immense variation in the quality of work on such inscriptions suggests that while some stonemasons might be capable of carving rough letters, that does not mean that they themselves are literate. It is not impossible that a carver's final product bears little resemblance to the message the customer demanded. A prime example of this is altars to the Celtic god Veteris, who seems to appear on altars in the north of Britain under such names as Deo Veteribus, Votri, Viteribus, Vitiribus, Hveteri, and Hvitri. If there can be such variation in one name, the potential for disastrous mistakes in translation elsewhere is huge.

We find similar issues with the *Notitia Dignitatum*, which seems to have lacunas and spelling errors. Moreover, it is entirely possible that the document as we know it is actually a composite of other documents, some of which considerably predate others. We do not have a date for the creation of the *Notitia*, though it is generally accepted that it dates from the opening decades of the fifth century. We also have no definite idea of its purpose, other than as a list of units and locations, cataloguing the forces serving under a number of officials across the empire. One fascinating theory is that it represents a grand Roman plan for retaking control of a fractured empire, and that would mean that the sections we deal with for Britain would involve a Roman strategy for the reconquest and reoccupation of the island following its abandonment in 410. Whatever the truth, it remains the latest documentary evidence for Roman Britain, and one of the most complete records available.

Britannia in a fifteenth century copy of the *Notitia Dignitatum* in the Bodleian. (*Creative Commons, Bodleian Library, Oxford*)

Part II

Roman Auxiliary Unit Listings

The pages that follow include listings for all known auxiliary units serving in Britain prior to the end of Roman government in AD 410. The order of each unit's inclusion is based not upon the numeral applied to the unit, but on its name, alphabetically. Each unit contains an in-depth description of their known or conjectural service record, constructed as far as possible from available evidence, followed by a list of known personnel and a catalogue of the epigraphy that mentions them.

The listings have been divided into three chapters. First come the *alae*, the cavalry units of the province, and at the end of the *ala* chapter, a sub-group has been included to allow a short entry for any *ala* about which we are uncertain, have insufficient data to deeply investigate, or are potential mistakes. This section is, again, alphabetical. The second main group concerns the cohorts, the infantry or mixed units of infantry and cavalry. Once again, they are listed alphabetically by unit name, regardless of number, and once again, they are followed by a sub-group of cohorts of uncertain status or insufficient evidence. The last group of entries concerns irregular units, either attested on inscriptions from the province or noted as present on the *Notitia Dignitatum*. The opening pages of each chapter provide a list of the units for ease of reference.

Throughout the listings I have favoured the modern names for sites mentioned, though I have given the ancient appellation where required. For personal names, when referencing personnel who are attested on inscriptions, I have tried to use the convention of 'i' instead of 'j' (for instance 'Iulius') in keeping with the epigraphy. However, when referencing known personalities in my discussions, I have used the more familiar j for ease, such as 'Julia Domna'.

Chapter 7

Alae

Agrippiana Miniata
I Hispanorum Asturum
II Asturum
Gallorum et Thracum
Gallorum Indiana
Gallorum Picentiana
Gallorum Proculeiana
Gallorum Sebosiana
I Herculiana
Nerviorum
Augusta Ob Virtutem Appellata
Gallorum Petriana
Pannoniorum Sabiniana
Sarmatarum
Pannoniorum Tampiana
I Thracum
I Tungrorum
Hispanorum Vettonum
Augusta Vocontiorum

Ala I Hispanorum Asturum (aka Ala Praetoria I Hispanorum Asturum)

The Ala I Asturum was a 500-strong cavalry wing raised in northern Spain, in what is now the region of Asturias. Their origin date is unknown, likely during the Augustan era, along with II Asturum, or possibly as a Neronian raising to provide extra cavalry regiments in the wake of the Boudiccan uprising. Our first potential evidence of them is the tombstone of an Ubian by the name of Albanus Excingi, who may have served with them in the first century before being buried in Chalons-sur-Saone (Cavillonum), although the unit attested is vague, and may instead refer to Ala II Asturum.

The unit is attested in Britain from AD 98 (diploma *CIL* XVI, 43). This means they may have been on the island since the invasion in AD 43, or more likely since 71 as part of the reinforcements brought to Britain in the wake of Boudicca's revolt, which would suggest that the unit may have been involved in putting down the Batavian revolt at the time of Excingi's service. Their garrison location between then and the first epigraphic evidence cannot be confirmed. The tombstone of a highly decorated former prefect found in Alcala del Rio, Spain (*ILS* 2712), names him as having served his military career with the *ala*, as well as with Cohors II Vasconum and Legio II Augusta, both of which had also been serving in Britain prior to 109, and it seems likely that he would have remained in the same province between positions, so that lends weight to an early British garrison for I Asturum. Nothing can be read into the order of the man's career given on the stone, as such inscriptions are not necessarily in chronological order.

Davies considers the possibility that the unit may be the one described by Tacitus in the Agricola as being destroyed by the Ordovices in 77, reinforcements being drawn in from Asturias, handily accounting for the double ethnic name, 'Spanish Asturians'. This is, of course, conjecture, but remains an intriguing possibility. If this is the case, that would put their Flavian garrison somewhere in North Wales.

The *ala*'s first (and only) garrison of which we can be certain is Benwell, on Hadrian's Wall. The earliest datable epigraphic evidence we have for the unit in Britain is the tombstone raised by a trooper named Numerianus to his freedman, a Mauri called Victor, at South Shields, dated to the second century. Despite its find spot, this should not be taken as evidence of the unit's presence there. It is likely that the unit were based at a local fort (Benwell is only 10 miles from South Shields), and that Numerianus retired to Arbeia and buried his freedman there after his period of service.

Given this, the unit may have garrisoned Benwell in the late second century. Moreover, the tombstone of Tineius Longus found at Benwell (*RIB* 1329) does not mention his unit, but marks him as a prefect of cavalry and has been tentatively dated to the reign of Commodus. This may be another indication that the unit had replaced Cohors I Vangionum at Benwell in the late second century.

The unit was certainly in Benwell by AD 205–208 (*RIB* 337), and still there under Gordian III in mid-AD 238, when their prefect Titus Agrippa restored a temple and raised an inscription to the mother goddesses of the parade ground and the genius of the unit. Ala I Asturum must have done something to distinguish themselves during the reigns of the Gordiani, for in that same inscription, they are named 'Gordian's Own'.

Tombstone of the Moor Victor, freedman of Numerianus, Arbeia Museum. (*Author's collection*)

The unit seems to have been based at that fort until its final occupation, for the *Notitia Dignitatum* lists, on the line of The Wall, '*Praefectus alae primae Asturum, Conderco*.' An undated inscription (*AE* 1940, 109) identifies the Ala II Asturum at Benwell, but this seems likely to be a mistake or some other aberration, perhaps a mis-carving or a lone retired trooper, as that unit seems likely to have been based at Ribchester in the second century and then Chesters.

The altar set up by the trooper named Victor at Saint-Géréon in France, dated to the end of the second or start of the third century (*AE* 1990, 732) seems likely to reference the Ala I Asturum Pia Fidelis, a different unit stationed largely in Germany and the Balkans, and so should not be taken as evidence of this unit being moved. The tile stamp from Wallsend cannot be taken as evidence of garrison presence. A fragmentary inscription unearthed in London records 'ALA I A', which could be this unit, but is undated and provides no confirmed evidence. An enigmatic fragment found in an unknown location on Hadrian's Wall names an 'N EGNATUS' as serving with what appears to be 'ASTURUM', so it is possible that this Gnaeus Egnatius was also one of the unit.

Dedication to the Mother Goddesses of the Parade-ground and the Genius of the unit from Benwell, now in the Hancock Museum. (*Author's collection*)

Attested officers and men:
Prefect Titus Agrippa, at Benwell, AD 238
Prefect Quintus Fulvius Euchir, from Alcala del Rio, Spain
Prefect Tineius Longus at Benwell, late second century
Trooper Numerianus, attested at South Shields in the second century
Victor, Mauri freedman of Numerianus at South Shields in the second century
Aurelius at Benwell, of uncertain date
Gnaeus Egnatius on Hadrian's Wall
Albanus Excingi at Chalons-sur-Marne, first century

Inscriptions:
Alcala Del Rio (Illipa Magna) – tombstone of Quintus Fulvius Euchir (*CIL* II, 1086)
Benwell (Condercum) – dedication March–June AD 238 (*RIB* 1334)
Benwell (Condercum) – dedication c. AD 208 (*RIB* 1337)
Benwell (Condercum) – unclassified and undated inscription (*RIB* 1348)
Benwell (Condercum) – tombstone of Aurelius (*RIB* 1349)
Benwell (Condercum) – tombstone of Tineius Longus (*RIB* 1329)
Benwell (Condercum) – stamped tiles (*RIB* 2464)
Chalons-sur-Saone (Cavillonum) – tombstone of Albanus Excingi (*CIL* XIII, 2613)
Efes (Ephesus) – inscription of a prefect of the *ala* (*ILS* 1366)
London (Londinium) – fragmentary inscription (*RIB* 02.08.2504.23)
South Shields (Arbeia) – tombstone of Victor (*RIB* 1064)
Wallsend (Segedunum) – stamped tile (*RIB* 2464)

Ala II Asturum Antoniniana

Our first evidence of the Ala II Asturum, which was likely created in northern Spain in the reign of Augustus, comes from a diploma of AD 61, in the reign of Nero (*AE* 1998, 1056), which places the unit at that time in Illyricum. It is likely during this period of service that a man named Receptus, presumably a retired veteran of the unit, died and was commemorated with a tombstone at Budapest. The unit garrisoned Dunaujváros (Intercisa) prior to AD 92, as confirmed by the first-century tombstone of Tiberius Claudius found there (*AE* 1992, 1458), following which it must have moved, perhaps straight to Britain, or alternatively via a brief occupation in Germany, if the tombstone of Albanus Excingi refers to this unit (see Ala I HIspanorum Asturum).

The first sign of the *ala* on British soil comes on a diploma from Brigetio in Hungary dated 122, perhaps of a man who had signed up in the region while the unit was still there and had been transferred to Britannia, only to return to Brigetio on his retirement. The evidence of the unit at Ribchester is undated, though it would seem to be the unit's first base on the island, sometime after the Neronian era, and perhaps under Agricola. The next unit attested at Ribchester (Ala Primae Sarmatorum) is transferred to Britain in AD 175, and so Ala II Asturum likely moved garrison at that time. As such, the famous Ribchester Helmet almost certainly belonged to a man of this unit. The tombstone of a decurion from the *ala* found at Lincoln is perfectly logical, as that city is a reasonable retirement place for a soldier from Ribchester.

Within a few short years we have our first evidence of the unit's next posting, at Chesters on Hadrian's Wall. Cohors I Vangionum had occupied this fort under Antoninus Pius but then moved to Risingham. Ala II Asturum is first attested between 178 and 184 in the dedication of part of the water system (*RIB* 1463) which dates their occupation to the governorship of Ulpius Marcellus. Inscriptions continue to attest the unit throughout the time of the Severans, with another building dedicated to the emperors Septimius Severus and Caracalla, and the caesar Geta, which dates occupation to AD 205–208. It seems likely that around this time the *ala* was involved in Severus' campaigns in Scotland and it may be a sign of some success and imperial favour that the unit acquired the epithet 'Antoniniana', referencing Caracalla by his more proper name. Though *RIB* 1049 from Chester-le-Street seems likely to reference the unit with the text 'EQ[...]NIANAE', given the fact that the unit is confirmed at Chesters by this time it is more likely that a visiting party from there helped build the Chester-le-Street aqueduct.

RIB 1470, a dedication by Aelius Longinus, mentions no unit explicitly, but he may well have been a member of this unit serving at Chesters. Similarly Marcus Aurelius Victor's unit is unnamed on *RIB* 1481, but the inscription has been convincingly dated to the third century, which would make him also a member of the unit.

The unit is still found in Chesters on 30 October AD 221, when the prefect Septimius Nilus dedicated a restored building in the fort to Emperor Elagabalus, and seems to have been based there for the rest of its service in Britain, for the *Notitia Dignitatum* notes the presence of '*Praefectus alae secundae Asturum, Cilurno.*'

Attested officers and men:

Prefect Septimius Nilus at Chesters, AD 221
Decurion Marcus Ingenuius Asiaticus, of uncertain date
Curator Turmae Aventinus, of uncertain date
Decurion Aelius Gemellus, son of Aventinus, of uncertain date
Decurion Marcus Ingenuius Asiaticus at Ribchester, of uncertain date

Decurion Titus Iulius Bellicus in the first century
Prefect Aelius Longinus? at Chesters, of uncertain date
Marcus Aurelius Victor at Chesters in the third century
Tiberius Claudius [...]onis in the first century
Receptus, son of Troucetissae in the first century

Inscriptions:
Benwell (Condercum) – fragmentary inscription (*AE* 1940, 109)
Braga (Bracara Augusta) – tombstone of Tiberius Claudius (*AE* 1992, 1458)
Budapest (Aquincum) – tombstone erected by Receptus, son of Troucetissae (*CIL* III, 14349.08)
Carlisle (Luguvalium) – lead seal (*RIB* 2411.83)
Chesters (Cilurnum) – dedication to Septimius Severus, Caracalla and Geta (*RIB* 1462)
Chesters (Cilurnum) – water tank dedication (*RIB* 1463)
Chesters (Cilurnum) – building dedication (*RIB* 1464)
Chesters (Cilurnum) – dedication to Elagabalus (*RIB* 1465)
Chesters (Cilurnum) – dedication to the Valour of the Emperors (*RIB* 1466)
Chesters (Cilurnum) – dedication by Aelius Longinus (*RIB* 1470)
Chesters (Cilurnum) – tombstone of Aventinus (*RIB* 1480)
Chesters (Cilurnum) – tombstone of Marcus Aurelius Victor (*RIB* 1481)
Chesters (Cilurnum) – lead seal (*RIB* 2411.17)
Corbridge (Coria) – lead seal (*RIB* 2411.82)
Gonobitz/Celeia – tombstone of Titus Iulius Bellicus (*CIL* III, 15205.3)
Lincoln (Lindum) – tombstone of a decurion (*RIB* 266)
Ribchester (Bremetenacum) – altar to the mother goddesses (*RIB* 586)

Ala Gallorum Indiana (aka Ala Gallorum Indiana Civium Romanorum)

The Ala Gallorum Indiana is a unit with an interesting history and interesting personnel, even if their stay in Britannia seems to have been relatively short. They are one of the slew of Gallic units raised in the first century and seemingly named after their first commander. We are told by Tacitus in his *Annals* (3.42) that:

> Julius Indus, a countryman of the insurgents, at feud with Florus and hence the more eager to be of service, was sent ahead with a body of picked men, and dispersed the still orderless multitude.

This strongly suggests that during the Sacrovir Revolt of AD 21, an irregular cavalry unit of Treveri, led by one Julius Indus, helped put down the revolt.

The Ribchester Helmet, British Museum. (*Author's collection*)

That the unit might then be incorporated into the standing Roman military as an *ala* bearing their leader's name makes perfect sense.

It is reasonable to assume a first garrison posting at Worms on the Upper Rhine, based upon the tombstone of Argiotalus Smertulitani, dated to AD 14–37, and we cannot be certain when the unit moved to Britain. There are various theories as to the date of this transfer. The conclusion that they took part in the initial invasion in 43 is tempting, given the dates, and their presence noted only as far north as Cirencester. Other historians favour a Neronian-era transfer, and others still a transfer under Vespasian. The presence of the name Flavius on the British inscription from Cirencester would strongly suggest a presence there during the Flavian era, though that name could also be reconstructed as Fulvius, and so is far from certain evidence. A diploma of AD 98 places the unit in Germania Inferior, and it is next attested at Worringen on the Lower Rhine in the late first century, and then at Mainz and nearby Klein Winterheim by the third century, as well as appearing on another diploma for Germania Superior in 134.

The only recorded prefect, whose name is sadly missing from the inscription, probably commanded the unit while in Germany, given that the vast majority of his career took place in Gaul and Germany. Similarly, Telesporus, recorded in Viterbo, is noted as a member of this unit, but also of Ala III Asturum, who were based in Germany, and so it seems reasonable to assume that Telesporus was also based with the unit there.

In terms of characters, as well as a prefect, decurions and troopers, we are given something of an insight into the array of ancillary staff such a unit would include among their garrison. Telesporus is recorded as a *medicus*, a medical professional, which suggests that the unit had a dedicated medical section, which would mean that its garrison forts would likely be equipped with a *valetudinarium*. We also have an inscription from Mainz to an unnamed *pequarius* (or *pecuarius*), a soldier with specific duties in the care of animals, surely an indispensable character in a cavalry unit.

The unit history, then for the Ala Gallorum Indiana, is fairly easy to reconstruct. That Germany is missing from the surviving copies of the *Notitia Dignitatum* makes it difficult to ascertain what units remained into the fifth century.

Attested officers and men:
Decurion Flavius Flavianus Aventinus at Mainz in the third century
Decurion Iulius Paterninus at Mainz in the third century
Decurion Albanus at Cirencester
Medicus Marcus Ulpius Sporus
Trooper Dannicus at Cirencester
Trooper Argiotalus Smertulitani at Worms AD 14–37
Trooper Fronto Dregenis of the Ubii at Mainz
Trooper Albinius Vitalis at Worringen late first century
Fulvius Natalis at Cirencester
Fulvius/Flavius Bitucus at Cirencester

Tombstone of Dannicus in Cirencester Museum. (*Author's collection*)

Inscriptions:
Cirencester (Corinium) – tombstone of Dannicus (*RIB* 108)
Klein Winterheim – Sarcophagus of Iulius Paterninum (*CIL* XIII, 7257)
Mainz (Mogontiacum) – inscription to a *pequarius* (*AE* 1940, 116)
Mainz (Mogontiacum) – inscription to Fronto Dregenis (*AE* 1929, 130)
Mainz (Mogontiacum) – inscription to Fl Flavianus Aventinus (*CIL* XIII, 7028)
Rome – Career inscription of an unnamed prefect (*CIL* VI, 1641)
Viterbo – inscription of Marcus Ulpius Sporus (*CIL* XI, 3007)
Worms (Borbetomagus) – tombstone of Argiotalus Smertulitani (*CIL* XIII, 6230)
Worringen (Burungum) – inscription to Albinius Vitalis (*CIL* XIII, 8519)

Ala Gallorum Picentiana

There is some suggestion that this unit was raised in Dalmatia, since a building inscription from Lesina (*CIL* III, 10094) names a cavalry prefect by the name of L. Rustius Picens. While it is a known phenomenon for units to take as their name that of their first commander, this is a process usually noted in Gaul and not Dalmatia, so a raising there seems unlikely, especially given the Gallorum element of the unit's name. It is most likely that the unit was raised by Picens in Gaul and that his presence in Lesina was a mark of him returning to his home after serving as the commander of the *ala*. Similarly the career inscription CIG 3991 from Konya in Turkey cannot be taken as evidence of the unit's presence there, but more likely of a retired commander.

Our first evidence for the unit itself, rather than Picens, comes from Germany, and of that there is plenty. Three tombstones attest their presence in Mainz, and one at Dienheim, just 10 miles upriver from there and with no recorded Roman presence of its own. In that rarest and most wonderful of all historical evidence, Tacitus (*Histories* 4.62) actually tells a story of the unit during the Civilis Revolt of AD 69:

> The squadron of Picentine horse could not endure the joy exhibited by the insulting mob, but, scorning the promises and threats of Sanctus, rode away to Mainz; on the way they happened to meet Longinus, the assassin of Vocula, whom they buried under a shower of weapons and so began the future expiation of their guilt.

The unit is not mentioned in the German diploma of AD 90, and therefore must be gone from the province before then. It is theorized that the *ala* was one of the many units shipped to Britain under the Flavians, either as part of the army of Petillius Cerialis in AD 71 or of Agricola in AD 77, though diploma evidence refutes this. The Debelets and Sikator diplomas, found in Thrace and Pannonia, are both understood to refer to units in Germany, and therefore place the Ala Picentiana there in AD 74 and 82. We therefore have only a narrow window for their departure from Germany, between AD 82 and 90. Since Domitian campaigned against the Chatti in 82/83 and drew troops from Britain to support his war, it seems likely that the *ala* was present for the Chatti War and then transferred to Britain in 83 or 84 when the war was over.

Whatever year they arrived, they are noted in diplomas of AD 122 and 124 as part of the military in Britain. Here, they are attested in only one place: the Flavian fort of Malton. They are the only known garrison of Malton from epigraphy, though here we run into a mystery. In the *Antonine Itinerary*, Route 1 runs from Corbridge via Ebchester, Binchester, Catterick, Aldborough and York, but then disintegrates into a journey involving three sites that are as yet unidentified: Derventio, Delgovicia and Praetorio. There have been many candidates put forward for these three sites over the years, with Malton often being identified with either Derventio (it stands on the Derwent) or Delgovicia. It can only be the Derventio of the *Antonine Itinerary* if there was a mistake in transcribing distances at some time. The identification of Praetorio with Brough-on-Humber (also known as Petuaria) is tempting. If Malton is to be identified with Derventio, then the unit has either gone by the time of the *Notitia Dignitatum*, or has assumed a new name, for the unit associated in that text is Numerus Supervenientium Petueriensium, suggesting a unit supervising the port at Petuaria. A more likely answer, given the large distance between there and Malton is that Derventio is one of the other sites, such as Stamford Bridge, closer to Petuaria, and that Malton's Roman name remains to be found. Unless epigraphic evidence turns up in an excavation, we will never know whether Malton is Derventio, and will therefore never know what happened to the *ala* based there.

Attested officers and men:
Prefect Candidus at Malton
Prefect L Rustius Picens at Lesina
Commander Leukion Pupion at Konya
Decurion Gaius Iulius at Mainz
Singularis Consularis Aurelius Macrinus at Malton
Trooper Silius Attonis at Dienheim
Trooper Abaius at Mainz
…us Paturi… at Mainz

Inscription from Malton, now in storage. (*Author's collection*)

Inscriptions:
Dienheim – tombstone of Silius Attonis (*CIL* XIII, 6277)
Hvar (Lesina) – inscription of L Rustius Picens (*CIL* III, 10094)
Konya (Iconium) – inscription of Neronian date (*CIG*, 3991)
Mainz (Mogontiacum) – tombstone of Gaius Iulius (*CIL*, 7296)
Mainz (Mogontiacum) – tombstone of Abaius (*AE* 1962, 290)
Mainz (Mogontiacum) – tombstone of Paturi (*AE* 1910, 211)
Malton (Derventio) – tombstone of Aurelius Macrinus (*RIB* 714)
Malton (Derventio) – dedication slab of Candidus (*RIB* 3207)

Ala Augusta Gallorum Proculeiana

A great deal of discussion surrounds both this unit and the Ala Augusta Ob Virtutem Appellata, with disagreement on connections between the two units and their origins. There is a strong belief that the Proculeiana and the Ob Virtutem Appellata are, in fact, the same unit, having changed its name, though again this is based solely upon speculation and divides historians. This notion is founded on the principle that the Proculeiana are mentioned specifically only on diplomas with no confirmed epigraphic evidence, while the Ob Virtutem

Appellata are well attested in epigraphy, but appear on no diploma. This is not sufficient evidence to say with any certainty that they are the same unit and so for clarity the two units have been treated separately in this text.

There are speculations that the Proculeiana, a quingenary cavalry unit, was first garrisoned in Germany. Its origin has been suggested as Gallia Lugdunensis, though again there is insufficient evidence to confirm that. What *can* be inferred with little doubt from the unit's name is that it was raised somewhere among the Gallic peoples, likely the Treveri (see below). Given the name it is likely their first commander was a man named Proculeius, conceivably Gaius Proculeius, a friend of Augustus.

The variety of diplomas that mention the unit confirm its presence in Britain from at least AD 98 to 146, and while no inscriptions overtly name this unit, two from Lancaster are highly likely candidates. These two cover the late first to early second century. The tombstone of Insus, now in Lancaster City Museum, tells us that he was a rider in a cavalry troop named Ala Augusta (of which there are several in Britain), but also tells us that he was of the Treveri, from around modern Trier. Similarly, the epitaph of Lucius Iulius Apollinaris confirms another Treveri in the service of an Ala Augusta. The Treveri being a Gallic tribe, the equating of these men with the Ala Augusta Gallorum seems highly likely, and thus lends a great deal of weight to the idea that Lancaster (possibly Calunium?) was the base for the Ala Augusta Gallorum Proculeiana at the turn of the first century. They seem to have moved elsewhere by the mid-second century, when the fort is garrisoned by the Ala Sebosiana. Their later service record is unknown.

Their last mention is AD 146, and while it is possible that the unit continued in service with a change of name, we can be sure of no service record after the mid-second century. For the potential that they did change and continue, see the entry for Ala Augusta Ob Virtutem Appellata.

Attested officers and men:

Curator Victor at Lancaster AD 75–120
Trooper Lucius Iulius Apollinaris at Lancaster
Insus son of Vodullus at Lancaster AD 75–120
Domitia, heir (wife?) of Insus at Lancaster AD 75–120
Propinquus (from the Walcot diploma) AD 126

Tombstone of Insus in Lancaster Museum. (*Author's collection*)

Inscriptions:
Lancaster (Calunium?) – tombstone of Insus (*RIB* 3185)
Lancaster (Calunium?) – tombstone of Lucius Iulius Apollinaris (*RIB* 606)

Ala Gallorum Sebosiana

The first evidence of the Ala Sebosiana comes from Germany, where they are attested at Worms and at Weisenau near Mainz in funerary inscriptions. It is likely that the unit was raised in the reign of Tiberius, possibly under the command of one Statius Sebosus, an acquaintance of Pliny's. The unit is unusual in being referenced directly in an ancient text, for Tacitus tells us in his *Histories* (III, 6) 'At Ateste they heard that three cohorts of Vitellian forces and the squadron of cavalry called Sebosian had occupied Forum Alieni and built a bridge over the stream there.' From this it is reasonable to assume, given that the unit was stationed in Germania superior and Vitellius was governor of Germania Inferior, that the Sebosiana threw in their lot with the new emperor along with the other German forces, as per Tacitus's suggestion.

It is highly unlikely that the unit would be returned to its posting when Vitellius fell and the Flavians gained the throne, and so it can be assumed that the Sebosiana were transferred to Britain along with other German units following the Civilis Revolt, under the command of Petillius Cerialis. Thus, we would find the Ala Gallorum Sebosiana in Britain from around AD 71. The unit is attested in Britain five times in diplomas between AD 103 and 178, but there is fascinating evidence to predate this.

One of the most unusual pieces of epigraphy that has come to light concerns this unit. A wooden writing tablet discovered at Carlisle, some of the earliest records of Roman Britain (A2395), names a rider of the Ala Sebosiana as being seconded to the bodyguard of Agricola: '[*eq (uiti) al*]*ae Sebosianae sing(ulari) uacat Agricolae uacat*.' While there is, indeed, more than one Agricola connected with Roman Britain, the dendrochronological dating of timbers in the early Carlisle fort to AD 72–73 and the understanding that Agricola was campaigning in the area as commander of Legio XX and sundry other forces, there is a strong suggestion that the Ala Sebosiana were the initial garrison of Carlisle during that first push northwards in the early 70s. Indeed, Tacitus tells us that Cerialis was 'putting him in command of part of the army, and sometimes, on the strength of the result, of larger forces.' The timber fort of Carlisle was first built in AD 72–73, but it was rebuilt AD 83–84, quite possibly by the same unit, since this would again be under Agricola, and it may have been then that the rider from the unit was seconded to Agricola's guard, since the man was governor at this point. Carlisle goes out of use at some point after 103, but in between, the Ala Petriana seems to have taken up residence, and so it is possible that Ala Sebosiana left Carlisle early, perhaps even accompanying Agricola to Mons Graupius in 83.

The *ala* must have been involved in the Antonine advance into Scotland. A tombstone found at Inveresk in 2007 clearly mentions them and has been dated to the Antonine period. Once the border was pulled back again, the unit must have returned south, and it seems likely that it was at this time they moved to Lancaster, replacing the Ala Augusta Gallorum Proculeiana. Second- to third-century evidence of the unit there abounds, including epigraphy, lead seals and stamped tiles. There is some suggestion from the context of tile finds that the unit was also in Lancaster in the Flavian era, but it is far from conclusive, and troop movements fit much better with their first arrival after 103, probably around AD 123 or 124.

The presence of a lead seal of the *ala* at Brough cannot be taken as evidence of their presence, and was most likely attached to a document sent from Lancaster. Similarly, the fascinating altar found on Bollihope Common near Binchester cannot be taken to prove their presence there, and is more suggestive of a visiting officer hunting in the area. Richmond marks the stone as third century or later, and Binchester's garrison at the turn of the third century was Ala Hispanorum Vettonum, so it is at least possible the Ala Sebosiana were in Binchester afterwards. The stone is worthy of note, as a snapshot of life in Roman Britain's northern military:

> To the Divinities of the Emperors and Unconquerable Silvanus Gaius Tetius Veturius Micianus, prefect of the Sebosian Cavalry Regiment, on fulfilment of his vow willingly set this up for taking a wild boar of remarkable fineness which many of his predecessors had been unable to bag.

The Lancaster altar to Ialonus Contrebis has raised a suggested unit link to Spain, given that Contrebia was a Celtiberian cult site in Hispania Tarraconensis, and that an Ala II Gallorum is referenced in the peninsula, but the presence of an altar to the same god in Nîmes has brought about a reassessment and a general acceptance that this is a Gallo-British Celtic deity.

We can certainly date to within a close period the inscription *RIB* 605. Since it names the rebel emperor Postumus (his name scratched out twice), it can be dated to between AD 260 and 269, and it appears to reference otherwise unknown consuls named Censor and Lepidus for a second time, which further limits the range to between AD 263 and 269. At this time, the unit acquires the honorific 'Postumus's Own', a title I imagine they dropped like a hot coal the moment the rebel empire fell and Rome regained control. The fate of the *ala* is unknown, for if the Ailione of the *Notitia Dignitatum* is to be equated with Lancaster, then the final unit posted there is one Cohors III Nerviorum, indicating that the Ala Sebosiana was gone by then. If the Bollihope Common altar dates to the late third century, as mentioned above it may indicate that this *ala* moved from Lancaster to Binchester, the nearest fort to the find site.

Attested officers and men:

Prefect Flavius Ammausius at Lancaster AD 263–269
Prefect Tetius Veturius Micianus, third century
Prefect Lucius Titius Manuetus
Decurion Iulius Januarius at Lancaster AD 241
Decurion Marcus Sempronius at Worms
Decurion Valerius at Brough-under-Stainmore
Trooper Crescens at Inveresk AD 240–260
Leubius Claupi at Worms
Tiberius Iulius Diviciacus at Mainz

Inscriptions:

Brough-under-Stainmore (Verteris) – lead seal of Valerius (*RIB* 2411.89)
Bollihope Common – altar to Silvanus (*RIB* 1041)
Carlisle (Luguvalium) – writing tablet (A2395)
Inveresk – tombstone of Crescens (*BRIT.* 39.5)

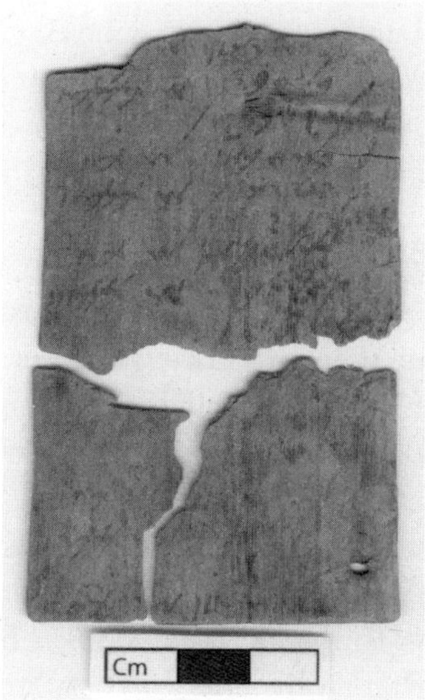

Carlisle tablet mentioning a rider of the ala, now in storage. (*Author's collection*)

Bathhouse dedication to Postumus in Lancaster Museum. (*Author's collection*)

Lancaster (Calunium?) – inscription to Postumus (*RIB* 605)
Lancaster (Calunium?) – altar to Ialonus Contrebis (*RIB* 600)
Lancaster (Calunium?) – lead seal (*RIB* 2411.88)
Lancaster (Calunium?) – stamped tiles (*RIB* 2465.1, Roman Britain 1973, Lancaster 25 and 26)
Rome – career inscription of L. Titius Mansuetus (*AE* 1971, 46)
Weisenau – (Finke 00216)
Worms (Borbetomagus) – (*CIL* XIII, 6236)
Worms (Borbetomagus) – (*CIL* XIII, 11709)

Ala Augusta Ob Virtutem Appellata

A great deal of discussion surrounds both this unit and the Ala Augusta Gallorum Proculeiana, with disagreement on connections between the two units and their origins. There are suggestions that Ob Virtutem Appellata were first garrisoned in Egypt, though evidence for this is inconclusive at best. The two Egyptian diplomas of 83 (*CIL* XVI, 29) and 105 (RMD I 9) mention only 'AUG' and could therefore refer to any one of a number of units bearing that honorific, not all of which were cavalry, and not all in Britain. There is a theory that the Proculeiana and the Ob Virtutem Appellata are, in fact, the same unit, having changed its name, though again this is based solely upon speculation, founded on the principle that the Proculeiana are mentioned only on diplomas with no epigraphic evidence, while the Ob Virtutem Appellata are well attested in epigraphy, but appear on no diploma. This is not sufficient evidence to say with any certainty that they are the same unit, though the fact that the last datable epigraphic evidence for Proculeiana is pre-Hadrianic, while the earliest for the Ob Virtutem *is* Hadrianic does lend some credence to the idea.

Almost all the epigraphy we have for the Ala Ob Virtutem comes from the fort of Old Carlisle, and we can be certain the unit was based there from at least AD 185 to AD 242, though they appear to have moved at some point in the third or fourth centuries, as the *Notitia Dignitatum* names the fort's final garrison as a Numerus Solensium. Of course, if the Ob Virtutem *is* the Ala Gallorum Proculeiana following a name change, it cannot be ruled out that a further change to Numerus Solensium occurred. *RIB* 897 from Old Carlisle, dated to AD 242, styles the unit 'Gordiana', though the reason for such imperial honour is unclear. An altar from Carlisle mentioning the unit and dated to AD 192 should not be taken as evidence of the unit's presence there, only 10 miles from their known base. Two other pieces of evidence for the unit have been unearthed at Chesters: a lead seal of uncertain date, and a Hadrianic altar. While the seal probably originated elsewhere, the fort of Old Carlisle seems to date from only mid- or late second century, and it is entirely possible that Chesters (also a cavalry fort) was the base for the Ob Virtutem before their move to Old Carlisle, being replaced by the Second Asturum. If so, and the unit is not to be equated with the Proculeiana, then they were perhaps brought to Britain during Hadrian's military reorganization.

Attested officers and men:
Prefect Septimenus Rusticus at Old Carlisle AD 185
Prefect Tiberius Claudius Justinus at Old Carlisle AD 188
Prefect Publius Aelius Magnus at Old Carlisle AD 191
Prefect Publius Sextanius at Carlisle AD 192
Prefect Egnatius Verecundus at Old Carlisle AD 197
Prefect Aemilius Crispinus at Old Carlisle AD 242
Prefect Rufinus at Old Carlisle
Decurion Lucius Valerius S[...] at Chesters
Sesquiplicarius trooper Maelonius Secundus at Old Carlisle
Curator Titus Aurelius Ateco at Old Carlisle
Latinianus, son of Rufinus at Old Carlisle

Inscriptions:
Carlisle (Luguvalium) – dedication to Hercules (*RIB* 946)
Chesters (Cilurnum) – altar to Hadrian (*RIB* 3298)
Chesters (Cilurnum) – lead seal of Valerius (*RIB* 2411.81)
Old Carlisle (Maglona) – altar to Bellona (*RIB* 890)
Old Carlisle (Maglona) – altar to Jupiter (*RIB* 893)
Old Carlisle (Maglona) – altar to Jupiter (*RIB* 894)
Old Carlisle (Maglona) – altar to Jupiter Dolichenus (*RIB* 895)
Old Carlisle (Maglona) – altar to Jupiter (*RIB* 897)
Old Carlisle (Maglona) – altar of Titus Aurelius Ateco (*RIB* 902)
Old Carlisle (Maglona) – dedication of Septimenus Rusticus (*RIB* 903)
Old Carlisle (Maglona) – dedication to Caracalla (*RIB* 905)
Old Carlisle (Maglona) – tombstone of Maelonius Secundus (*RIB* 907)

Dedication to Hercules from Carlisle, Tullie House Museum. (*Author's collection*)

Ala Gallorum Petriana (aka Ala Augusta Gallorum Petriana Milliaria Civium Romanorum Bis Torquata)

An interesting unit with a complex name, some of the best surviving evidence in the form of a glorious tombstone in Hexham Abbey, and a mystery concerning their service record. There is evidence in Germany, at Bonn (Bonna) of a unit which may at first seem unrelated. The tombstone there of one Niger Aetonis, dated on a stylistic basis to AD 43–50, names him as a member of a unit named Ala Pomponiana. This unit appears on no list or diploma after AD 74 and seems to have disbanded or changed name by then, which gives us an intriguing possibility. It is possible that the Ala Pomponiana became Ala Petriana in its early days in Bonn, between AD 17 and 56. This is an attractive idea, given the inscription of a man from Reggio Emilia by the name of Titus Pomponius Petrus, who is recorded as having served as a cavalry prefect, likely under Germanicus around AD 11–17. The *ala* may, therefore, have changed the way it honoured their commander's name from Pomponiana (Pomponius) to Petriana (Petrus).

We have evidence that the Ala Petriana existed in Germany by AD 56 as a quingenary cavalry unit based at Mainz (Mogontiacum). A statue base there, dated convincingly to 56, was raised in honour of Nero by one C. Iulius Augurinus, prefect of the Ala Gallorum Petriana (*ILS* 2491). Interestingly, this Iulius Augurinus is almost certainly the same Iulius Augurinus who was executed for his part in the Pisonian Conspiracy of AD 65, written of by Tacitus (*Annals* 15.50). The unit may have been created in the reign of Tiberius (as per Ala Pomponiana above), a time of great military activity in the region. The fact that it had acquired the epithet 'Gallorum' by AD 56 tells us that the unit was raised in Gaul, probably from the area of Trier, given a dedication by Fittius from Strasbourg (Argentoratum) also references this unit as Ala Petriana Treverorum (*CIL* XIII, 11605)

The next information we have for the unit dates to April AD 69, and comes from Tacitus (*Annals* 1.70):

> [Caecina] sent in advance infantry, made up of Gauls, Lusitanians, and Britons, and some German detachments with the squadron of Petra's horse.

What happened following the civil war of AD 69 is the mystery. We have two potential pieces of evidence for the *ala*, both in different regions, which presents us with two possibilities. The first is from Latschach in Austria (Roman Noricum): the tombstone of the decurion Sentius Niger, who retired to Virunum and predates AD 122. The second is the tombstone of the *signifer* Flavinus from Corbridge, which dates to the late first century. It is possible that the unit was transferred to the Danube, perhaps by Domitian, where Sentius Niger served in Noricum, and that the tombstone of Flavinus in Hexham dates from circa AD 98 when the unit was in Britain. It is more likely, however, based on unit titles, that the unit was moved straight to Britain post 69, based at Corbridge or the earlier Corbridge Red House fort, and that Sentius Niger retired from there to Noricum, where he was buried. By 98 the unit had acquired the title Torquata, which is not mentioned on the tombstone of Flavinus, but is on the that of Niger, suggesting that in sequence Flavinus predates Niger, lending weight to the Britannia alternative over the Noricum one.

This would mean that the unit was moved to Britain in the early Flavian era, perhaps under Cerialis in his campaign against the Brigantes, and most likely served with Agricola during the conquest of the north. It is therefore likely that the unit occupied the Red House fort at Corbridge under Domitian, and acquired its first torc for valour at Mons Graupius. By the time of Trajan's reign and the diploma dated 98 (*CIL* XVI, 43), the unit is confirmed as in Britain, and was likely in Carlisle at that time, where *RIB* 957 notes them as having won only one torc so far. This inscription also records the unit as '*Civium Romanorum*', suggesting they won their citizenship under Domitian also. At some point between 98 and 117, the *ala* expanded and became a 1,000-strong milliary cavalry unit, the only one in Britain. It may be that this change also coincides with the unit's name changing from Torquata to Bis Torquata, indicating the acquisition of a second torc for valour, which it certainly had by AD 122. This second award may be connected with the British revolts under governor Pompeius Falco recorded in coins of AD 119.

At around AD 160 the western turf stretch of Hadrian's Wall was replaced in stone, and at this time the fort at Stanwix (Uxellodunum) experienced a similar rebuild and was enlarged from a small infantry fort to house a milliary cavalry unit. It is almost certain that the Ala Petriana's occupation of Stanwix began then. What remains to be answered is the disposition of the unit between being expanded into a milliary force prior to AD 118, and their placement as a garrison at the large Stanwix site in AD 160, there being four decades of occupation between, during which 1,000 men would be stationed in Carlisle, a fort at least an acre smaller than Stanwix. It seems likely, as Bishop noted, that the force was split for a time between two forts. It is tempting to suggest that while much of the *ala* was based in Carlisle at this time, the rest occupied the Stanegate fort of Brampton Old Church (whose garrison has never been identified), moving from there to nearby Castlesteads when the wall was built. They could then easily have left Castlesteads and moved into the new fort of Stanwix around AD 160, when cohors IV Gallorum returned from Scotland and took up the Castlesteads posting. Though there is no proof of this, there may be one other indicator. *RIB* 2383 is an inscription carved into the rocks above Banks Burn, near Lanercost. Though it is usually considered not to be genuine, it names the Ala Petriana, and is close to both Brampton Old Church and Castlesteads. An intriguing possibility.

The unit seems to have remained at Stanwix throughout the rest of the Roman occupation, its presence in the province confirmed by diplomas of 122 (*CIL* XVI, 69) 124 (*CIL* XVI, 70) and 135 (*CIL* XVI, 82), and they were still the garrison by the time of the *Notitia Dignitatum*, which records: '*Praefectus alae Petrianae, at Petrianis.*' The fort of Stanwix is still named Uxellodunum in the Ravenna Cosmography, and so its inclusion in the *Notitia* as Petrianis is mystery, though its position between the neighbouring installation confirms that it *is* Stanwix. Perhaps the fort acquired the term Petrianis after its occupants sometime in the third or fourth centuries.

It is worth noting that none of the Vindolanda tablets referencing the unit should be take as evidence of their presence at the fort, as they all originated elsewhere and were sent *to* Vindolanda. Since the tablets are generally dated to the latter part of the first century and the early second, we can presume that the letters referencing the unit were sent from Corbridge.

Attested officers and men:
Prefect Titus Pomponius Petrus AD 11–17
Prefect Gaius Iulius Augurinus at Mainz AD 56
Prefect Luca at Stanwix
Prefect Gaius Camurius Clemens in early second century
Decurion Sentius Niger at Latschach
Decurion Candidus at Corbridge in late first century
Decurion Junius Brutus late first or early second century
Decurion Cluvius Florus between AD 97 and AD 105
Standard Bearer Flavinus at Corbridge in late first century
Veteran [...]gadunus from Xanten (Ulpia Traiana)
Martius, son of [...]gadunus
Fittius at Strasburg
Niger Aetonis at Bonn AD 43–50

Inscriptions:
Attidium – inscription of Camurius Clemens (*ILS* 2728)
Bonn (Bonna) – inscription of Niger Aetonis (*CIL* XIII, 8097)
Carlisle (Luguvalium) – dedication of Luca (*RIB* 957)
Chesterholm (Vindolanda) – letter to Flavius Cerialis (Vindolanda Tablet 258)
Chesterholm (Vindolanda) – letter to Flavius Cerialis (Vindolanda Tablet 277)

Tombstone of Flavinus in Hexham Abbey, probably from Corbridge. (*Author's collection*)

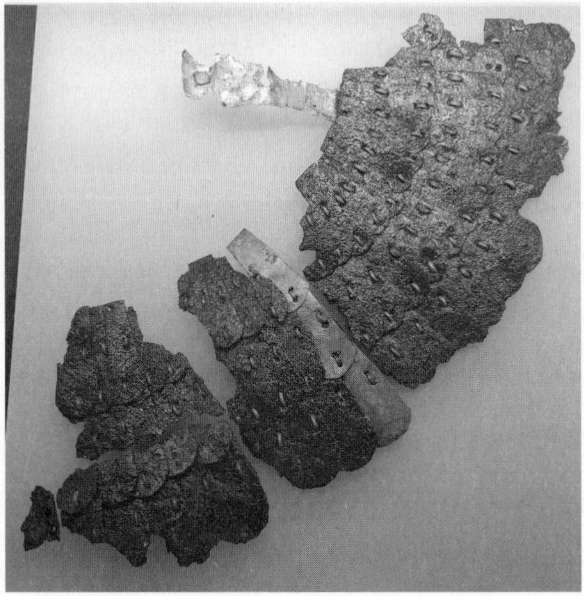

Lorica squamata from Carlisle, Tullie House Museum. (*Author's collection*)

Chesterholm (Vindolanda) – letter to the VIIII Batavorum (Vindolanda Tablet 281)
Chesterholm (Vindolanda) – letter (Vindolanda Tablet 700)
Hexham Abbey – tombstone of Flavinus (*RIB* 1172)
Lanercost area – quarry inscription of Iunius Brutus (*RIB* 2383)
Mainz (Mogontiacum) – inscription of Gaius Iulius Augurinus (IL 2491)
Old Penrith (Voreda) – tombstone of [...]gadunus (*RIB* 935)
Reggio Emilia – career inscription of T Pomponius Petrus (*CIL* XI, 969)
Stanwix (Uxellodunum) – lead seal (*RIB* 2411.84)

Ala I Pannoniorum Sabiniana

The origins of this *ala* are uncertain. The association of the unit with the infamous C. Nymphidius Sabinus (later Nero's praetorian prefect) based on *CIL* III, 4269 is uncertain, but tempting. Nymphidius Sabinus was in Pannonia under Caligula or Claudius (as altar *CIL* III, 4269, raised by him, is dated AD 41–54), and it is therefore quite possible that he commanded a newly raised Pannonian *ala*, giving his name to it, and then moved away with it to its next posting. Certainly, the unit will not have been created until after the Pannonian revolt of AD 6.

The unit's next mention and possible posting is given by the tombstone of one Aemilius Balaesus in Aldeia Nova (in modern Portugal). The inscription is dated to the Flavian era, which gives us the possibility that it formed the local Hispanic garrison. While Spain is generally seen as a peaceful province after the Cantabrian wars, there was an ongoing military presence, particularly in the north and west, where gold mines flourished.

When the unit came to Britain is also uncertain, but given the Flavian date of the Aldeia Nova inscription, it is likely to have been at the very earliest at the end of the first century. The unit's first dated attestation in Britain is a diploma of AD 122, with other diplomas confirming its presence in 145 and 178. As such, while it is possible that the unit moved to Britain in the late Flavian era, perhaps under Agricola, or under Nerva or Trajan, the most likely situation is that the unit was drawn from relatively peaceful Hispania by Hadrian to form part of the garrison of his new northern frontier.

The presence of stamped tiles and lead seals has led to an assumption that the *ala* was the first garrison of South Shields, but it is more likely that these minor pieces of evidence are the result of visits, communications, and possibly work parties from another site, and are therefore no guarantee of occupation. Similarly, seals found at Binchester and Corbridge cannot be taken as evidence of the unit's presence there, and all these seems likely to have originated at one central place.

The only real evidence of garrison occupation we have for the unit comes from the fort of Halton Chesters on Hadrian's Wall. It is an early third-century tombstone, whose subject remains unknown, but which we know from the inscription was raised by the man's brother, who served with this *ala*. As such, it is likely they both did, and therefore that this *ala* was based at Halton Chesters from the days the wall was constructed. We know they were still there at the turn of the fifth century, for the *Notitia Dignitatum* tells us of a '*Praefectus alae Sabinianae, at Hunno*.'

As an interesting aside, a tombstone from Rome (*CIL* VI, 3249) records a man drawn probably from this unit into the imperial cavalry bodyguard: '*Dis Manibus Cornelius Cornelianus eques singularis Augusti adlectus ex ala ... iniana*.'

Attested officers and men:
Signifer Aemilius Balaesus
Duplicarius ... [M]essorius Magnus, third century
Cornelius Cornelianus

Remains of *lorica hamata* from South Shields, Arbeia Museum. (*Author's collection*)

Inscriptions:
Aldeia Nova – tombstone of Balaesus (EE viii 128)
Binchester (Vinovia) – seal (*RIB* 2411.85)
Corbridge (Coria) – seal (*RIB* 2411.87)
Halton Chesters (Onnum) – tombstone (*RIB* 1433)
South Shields (Arbeia) – stamped tile (EE iii 202)
South Shields (Arbeia) – seal (*RIB* 2411.86)
Rome – tombstone of Cornelianus (*CIL* VI, 3249)

Ala Sarmatarum

What could have been one of the shortest entries in this text, due to minimal data, turns into one of the most interesting because of the circumstances and details surrounding it. In his *Roman History* (72.16), Cassius Dio tells us of Marcus Aurelius that the Iazyges 'promptly furnished as their contribution to the alliance eight thousand cavalry, fifty-five hundred of whom he sent to Britian'. The Iazyges are one of the tribes of a confederation called the Sarmatians who, as noted earlier in the text, occupied the Russian steppe and gradually migrated west until they encountered Rome in the region of what is now Romania. The Sarmatians were riders, horse people akin to the many other similar peoples who came from the East.

That figure of 5,500 riders is a huge amount of cavalry, even if it included the women, who would not be included in Roman garrisons. The largest standard auxiliary cavalry unit would number 1,000, and most would be 500 men. As such, the recruits sent to Britain by Marcus Aurelius (if Dio is telling us the truth) would form several auxiliary *alae*, and where the majority went remains unknown.

One possibility is that a number of men, as well as the women, were allowed to settle as civilians. It would make sense that a people so connected with horses might be used to breed them, train them, create cavalry equipment, and the like. Even so, it seems unlikely that near 5,000 such people were settled. It may be that many of those riders were split between extant under-strength cavalry units to make up the numbers, taking on the cultural identity of the units they joined. Only in one place is there confirmed evidence that those very Sarmatians remained and held to their ethnicity, though a case may be made for two more.

The Ala II Asturum left Ribchester and was installed at Chesters by around AD 178, which suggests strongly that they moved upon the arrival there of Aurelius' Sarmatians in 175. We know the Ala Sarmatarum was settled thereafter at Ribchester, based on three inscriptions, one of them dated to AD 241. Indeed, the *ala* remains in garrison there until the end of Roman rule in Britain, as the *Notitia Dignitatum* mentions the existence of '*Cuneus Sarmatarum Bremetenraco*'. In this respect the unit is unique in Britain in having changed form from an *ala* of the Imperial era to an equivalent *cuneus* of the late Roman period, a designation, ironically, that might have come down from the Huns, another horse people of the steppe who followed on in the wake of the Sarmatians.

It is possible that the cavalry tombstone found at Chester and now in the Grosvenor Museum, as well as fragments of horse armour, beads and other small finds along Hadrian's Wall suggested to have a connection with the Sarmatians, are evidence of other men from those 5,500 having been filtered into other units. For possible evidence of their presence elsewhere, see also Numerus Equitum Sarmatarum and Equites Catafractariorum.

As a final note, there is a repeated tendency to link this unit and its fellows brought over by Aurelius to the legend of King Arthur. Though there is no direct confirmation of a connection, the circumstantial evidence is intriguing and continues to foster theories.

Attested officers and men:
Prefect Aelius Antoninus at Ribchester AD 241
Singularis Consularis Iulius Maximus at Ribchester

Tombstone of a draconarius from Chester, Grosvenor Museum. (*Author's collection*)

Tamga symbol, probably of Sarmatian origin, found at Ribchester. (*Image from signboard at Ribchester Museum*)

Inscriptions:
Ribchester (Bremetenacum) – altar to Apollo (*RIB* 583)
Ribchester (Bremetenacum) – tombstone of Aelia Matrona and Marcus Iulius Maximus (*RIB* 594)
Ribchester (Bremetenacum) – decurion's tombstone (*RIB* 595)

Ala I Thracum

The Thracian Revolt of AD 26 was partly caused by their refusal to provide new troops for Rome, because such soldiers were being sent to far off locations, and it is possible that this *ala* is one of those very units. It is likely that Ala I Thracum was raised under Augustus, and there is a common belief that the unit was first based in Germania. The first records we have for the unit in Britain are the tombstones found in Cirencester and Colchester, both of which are assumed to date from the early days of the invasion, and so it is extremely likely that the unit was involved in the invasion in AD 43.

The *ala*, then, was probably in Britain from the mid-first century and was involved in the conquest from that time on. They were still recorded in Britain on the Malpas and Stannington diplomas of AD 103 and 124, but it seems very likely that it was just after this that they left the island. Inscriptions from Utrecht and Köln place the unit there by AD 150 at the latest, and so they may have been moved back to Germany during the Hadrianic reorganization of the provincial military, perhaps on the completion of his wall in 128. We cannot be certain where the unit was based at any time in Britain, as the tombstones found in the province were likely in locations to which the deceased had retired.

Attested officers and men:
Prefect Marcus Aurelius Hermogenes
Prefect Quintus Didius Lemonius Euhodianus at Köln AD 131–170
Duplicarius Longinus Sdape in Britain[1]
Trooper Sextus Valerius Genialis in Britain
Valentius Bititralis at Utrecht AD 117–150

Inscriptions:
Amiternum – career inscription of Ignotus (*AE* 1983, 325)
Ostia – career inscription of Hermogenes (*CIL* XIV, 5340)
Cirencester (Corinium) – tombstone of Genialis (*RIB* 109)
Colchester (Camulodunum) – tombstone of Longinus (*RIB* 201)
Caerleon (Isca) – trulla with maker's mark (*RIB* 2415.39)
Köln (Colonia) – tombstone of Euhodianus (*CIL* XIII, 12058)
Utrecht (Traiectum) – tombstone of Bititralis (*CIL* XIII, 8818)
Xanten (Vetera) – fragment (*CIL* XIII, 8659)

Tombstone of Longinus in Colchester Castle. (*Author's collection*)

Ala Hispanorum Vettonum (aka Ala Hispanorum Vettonum Civium Romanorum)

The Vettones were a tribe of central Iberia and were included within the empire early in the reign of Augustus. As such, the unit may have been raised there at

1. The full inscription for *RIB* 201 begins:
LONGINVS SDAPEZE
MATYGI F DVPLICARIVS
Since neither Sdapeze nor Sdapezematygi are considered Thracian names, it has been conjectured by both Seure and Detschew that there should be a break in the name on the first line, conjoining the ZE with the second, which would give the translation as LONGINUS SDAPE SON OF ZEMAYGI. This is the translation for which this author has opted.

any point from the time of Augustus onwards. We know that they were present in Britain from at least AD 103 thanks to the Malpas diploma, and remained in the province until at least AD 178 (again from a diploma), though the tombstone from Bath (*RIB* 159) has been given a Flavian or even Neronian date, suggesting the unit may have been in Britain since the invasion.

During their time they occupied two sites, attested by epigraphy: Brecon Gaer and Binchester. Though we have dates for only one inscription, it seems likely that Brecon Gaer was the first, the style of the stone likely dating it to late first or early second century, and Binchester the second of their two garrison sites, the unit being recorded at nearby Bowes helping the local garrison with building work in AD 197–198. The funerary inscriptions found in Bath and at Segermes in Tunisia are likely to be locations chosen by retirees, and so cannot be reasoned to be garrison sites. The fate of the unit after AD 198 is unknown.

Attested officers and men:
Prefect [...]onius Rufus
Prefect Valerius Fronto AD 197–198
Prefect Marcus Valerius Fulvianus at Binchester
Medicus Marcus Aurelius [...]ocomas at Binchester
Architectus Eltaominus at Binchester
Decurion Clemens at Brecon Gaer
Trooper Bassus Turobei
Trooper Lucius Vitellius Tancinus
Trooper Candidus at Brecon Gaer

Tombstone of Tancinus from Bath, Roman Bath Museum. (*Author's collection*)

The bath house of Binchester. (*Author's collection*)

Inscriptions:
Bath (Aquae Sulis) – tombstone of Tancinus (*RIB* 159)
Binchester (Vinovia) – altar of Rufus (*RIB* 1991, 10)
Binchester (Vinovia) – altar of Fulvianus (*RIB* 1029)
Binchester (Vinovia) – altar of Aurelius (*RIB* 1028)
Binchester (Vinovia) – altar (*RIB* 1032)
Binchester (Vinovia) – altar of Eltaominus (*BRIT*. 45.4)
Bowes (Lavatris) – altar of Fronto (*RIB* 730)
Brecon Gaer (Cicucium) – tombstone of Candidus (*RIB* 403)
Calerula – career inscription of Bassus Turobei (*AE* 1986, 426)
Henchir al Harat (Segermes) – career inscription (*AE* 1905, 128)

Ala Augusta Vocontiorum (aka Ala Augusta Vocontiorum Civium Romanorum)

Raised among the Vocontii of southeastern France, the wealth of locations referencing this unit makes its history somewhat cloudy. A seemingly early tombstone from Pontarcher near Soissons, without the Augusta honorific, suggests that it was based at the first century Julio-Claudian fort there. What its role could be in such a peaceful, settled region remains unclear. The Pontarcher fort stands at a Roman crossroads and at an important crossing of the Aisne, and so the garrison (fellow Gauls, notably) may have been there more in the role of police than in a military manner. Their presence was likely short-lived, given the scarcity of evidence, and it seems likely that the unit was then moved to Germania Inferior, probably in the Flavian era and perhaps for Domitian's war against the Chatti.

The unit's history in Germany is complex, and seems to have involved two periods. The Monterburg tombstones are dated between pre-Flavian (8671) and Flavian/Trajanic (8672), giving us a reasonable date for their presence. Members of the unit were clearly recruited in the area, given that Silvanus Loupi is of the Treveri, and the career inscription found at Mersch suggests the unit's presence in Germania Inferior in the very early second century. We cannot say for sure when the unit was transferred from Germany to Britain, but we have a certain window, which presents interesting questions. The unit was not mentioned in the Malpas diploma of AD 103, but is present on the Brompton diploma of AD 122. It must therefore have been moved to Britain between those years, perhaps by Trajan following the conclusion of the Dacian Wars in AD 106. The unit remained in Britain until at least AD 178 as attested by other diplomas, as well as the Newstead inscription, which places it there in the reign of Antoninus Pius. By the time of the 122 diploma, the unit had acquired its Augusta honorific which it bears on all subsequent diplomas as well as the Newstead altar, perhaps having been 'Domitiana' following the Chatti Wars, changed to Augusta following his *damnatio*.

A huge question is raised by the statue base found at Hemmen in Germania Inferior (modern Netherlands). The inscription refers to the Ala Vocontiorum, without its Augusta honorific, suggesting a date prior to AD 122. However, the inscription refers to the unit as Exercituus Britannici, suggesting that its presence in Hemmen was a temporary assignment for a vexillation of the unit while based in Britain. The problem is eased if we simply accept that the decurion Simplicius Super did not bother adding the honorific to his unit's title, though this seems unlikely given the effort he went to in detailing their British origin. Thus, we are left with the only realistic possibility being that the unit was moved to Britain from Germany not long after 103 in the reign of Trajan, but a vexillation was then sent back to Germany in the following decade or so, still under Trajan, where a decurion raised a statue in Hemmen. It is tempting to see the return of part of the unit to Germany as a response to Trajan's Parthian War of AD 113, perhaps bolstering the Rhine garrisons while other units were away at the war.

The lead seal at Leicester and the stamped tile from London cannot be taken as evidence of the unit's presence there as a garrison. We have no datable evidence for the activity of the unit following the diploma of AD 178, at which point it was still in Britain.

Attested officers and men:

Decurion Aelius Marcus at Newstead AD 140–155
Decurion Simplicius Super at Hemmen
Decurion Flavius Similis at Leicester
Trooper […]us Secci at Pontarcher
Trooper Silvanus Loupi at Köln
Atil[…]us Di[…]xti at Monterburg

Inscriptions:

Hemmen – statue base or altar of Simplicius (*ILS* 2536)
Köln (Colonia) – tombstone of Silvanus Loupi (*CIL* XIII, 8655)
Leicester (Ratae) – lead seal of Similis (*RIB* 2411.90)
London (Londinium) – stamped tile (EE iv 698)
Mersch (Marisca) – career inscription (*CIL* XIII, 4030)
Monterburg (Burginatium) – tombstone of Atillus (*CIL* XIII, 8671)
Monterburg (Burginatium) – tombstone (*CIL* XIII, 8672)
Newstead (Trimontium) – altar of Aelius Marcus (*RIB* 2121)
Pontarcher – tombstone of Secci (*CIL* XIII, 3463)

Decorative cavalry helmet from Newstead, now in National Museum of Scotland. (*Author's collection*)

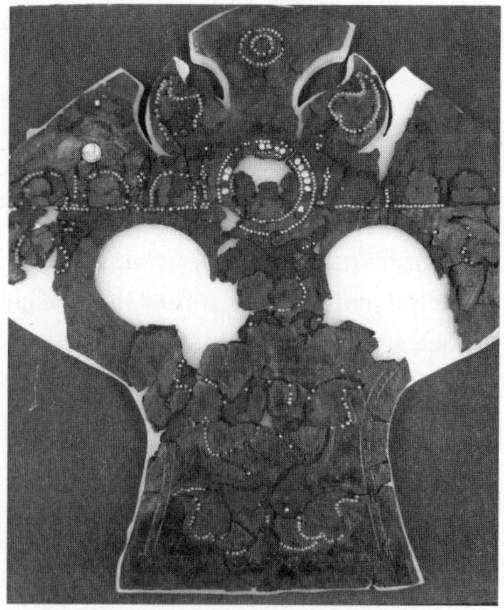

Leather horse chamfron from Newstead, now in National Museum of Scotland. (*Author's collection*)

ALAE OF UNCERTAIN STATUS OR SCARCE DATA

Ala Agrippiana Miniata

This unit, a cavalry wing, seems to be attested early on in Germany, where it may have attained its name (though see below). The gravestone of one Decimus Decmanius Caper, found in Grenoble and dated to before AD 50, names him as a subprefect of an Ala Agrippiana (*CIL* XII, 2231). Though the unit is attested on no extant inscription in Britain, the *ala* was garrisoned somewhere on the island during the reign of Hadrian, as well as possibly before and after. A military diploma (*CIL* XVI, 69) labels a list of units based in Britain under Aulus Platorius Nepos in 122, and though we have no other evidence for the unit, they are named on the British roll for retirement in that year. With regard to their name, 'Miniata' can be translated as 'Painted Red', while 'Agrippiana' may be named for one Lucius Nasidienus Agrippa, a commander mentioned in an inscription from Cologne (*CIL* XIII, 8270.) There are numerous problems with this unit's identification, due to lack of archaeological evidence, and it may or may not be the same unit as an Ala II Flavia Agrippiana which served in Parthia. If this is the case, then it is possible the name Agrippiana was acquired from Herod Agrippa II and not an early commander at all. If the various Agrippiana units are assumed to be variations in name on the same *ala*, then they would likely have been raised in Gaul, served in Germany and then Syria, been moved to Britain by the time of Hadrian, losing their Flavian epithet by the time of the Hadrianic diploma, and then back to Syria for post-Hadrianic service. No known sites or personnel are recorded for the unit in Britain.

Ala Gallorum et Thracum Classiana

This unit may have Augustan roots as a cavalry wing raised from Gaul, later bolstered with a strong influx of Thracian recruits, the unit renamed appropriately. Thrace was conquered in the early first century AD, and mass recruiting into the Roman *auxilia* is recorded after the fall of the Odryssian kingdom under Claudius in AD 46, and so it seems highly possible that the unit acquired its double moniker at that time. Classiana does not refer to a naval force, but rather points to the unit falling under the command of an officer by the name of Classius, whose name then became attached to the wing. This almost certainly all happened within the first century. It seems to have been a quingenary *ala* of 500, and its active service prior to arrival in Britain is unknown. It may have been assigned to the island shortly after the invasion, during one of the series of pushes north and west, perhaps as one of the units brought to the island by Petillius Cerialis. All we can say for certain is that the unit was based in Britain in the reign of Trajan, as they are recorded in the province in a diploma of 105. With soldiers being retired after twenty-five years of service, this confirms that the unit was in existence prior to AD 80. Their service seems to have been unusual, for by the reign of Hadrian the unit is attested in Germany, in a diploma of 127, but is once again back in Britain by the time of a third diploma dated 178. It may be that the unit occupied one of the forts north of the line of Hadrian's Wall and in 122 when the emperor was visiting the province and reorganized the army, the fort was abandoned, the unit withdrawn and then sent to a new posting in Germany. It is purely conjecture, but if this was the case, it is equally possible that when his successor, Antoninus Pius, reoccupied the north and advanced territory to the line of his new wall, the unit was brought back to Britain as an *ala* familiar with the terrain, possibly even sent to reoccupy its former garrison. The fate of the unit is unknown, and until any epigraphic evidence is unearthed, we are unlikely to be able to do more than conjecture as to their history.

Ala I Herculeana

This unit, the 'Herculean Cavalry Wing' is only attested in Britain in one source. The *Notitia Dignitatum* places the unit *per lineam Valli*, which is 'along the line of the wall' at an unknown station (*praefectus alae primae Herculeae, Olenaco*). Despite the heading, this section of the *Notitia* covers not only the wall forts, but a sequence of other sites thereafter, generally heading south. The Olenacum fort has been tentatively identified with Elslack, near Skipton, and the positioning in the *Notitia* between the entries for Ribchester and Brough-by-Bainbridge supports that suggestion. The unit's name is interesting and may give us a date for its creation. Emperor Maximian adopted the title 'Herculius' on his accession in 286, and so this unit may well have been either raised or adopted and renamed by the emperor in that year, possibly somewhere in Germany or Gaul. They may have come to Britain in 289 during Maximian's first abortive campaign against the usurper Carausius, but could not realistically have formed a garrison here until the island was retaken by Constantius in 296. Therefore it is possible to tentatively reconstruct the unit's post-Maximianic history. Between 286 and 289 they were either raised, or renamed from another older appellation, in honour of Maximian, and likely served in his campaigns on the German frontier. They may have formed part of the army of Constantius in 296, overcoming the usurper and returning Britain to imperial control. They would then have been settled into garrison in the newly recovered province at Elslack, where they then remained until the end of Roman government on the island. There is another known Thracian unit, the Ala Thracum Herculania, which predates the Dominate period, and may have acquired its name from an early commander, perhaps Herculanus or Gaius Julilus Heracles. There is no evidence to link the two units other than similarity of the name, but if they were to be considered together, that would place the unit in the East, in Syria in the second century and Egypt in the third before its transfer to Britain.

Ala Nerviorum

This unit is only positively referenced once, with a possible second inscription from the same location. It seems they were the garrison unit of the fort of Cicucium (Brecon Gaer), where two tombstones have been found (*RIB* 404 and 405), one, at least, datable to between the Flavian and Hadrianic eras. The latter refers to Valerius Primus, a rider of this unit, while the former, bearing the worn text 'AL … N…' likely references the same *ala*. There is no other record of their service, though the Nervii serving in Britain is well attested in infantry cohorts.

Ala Pannoniorum Tampiana Victrix

This *ala* is only referenced on two diplomas for Britain, most evidence of its existence coming from Noricum. Their dates in Britain are uncertain, but they were here in AD 103 and 122, according to the Malpas and Brigetio diplomas, and have moved to Noricum by the 130s. Their garrison and activity in Britain remains unknown. The Victrix epithet seems to postdate their time in Britain.

Ala I Tungrorum

We know practically nothing about this unit. Their origins are obscure, although undoubtedly raised around Tongeren, and while the evidence for their existence in Britannia is confirmed by numerous diplomas between AD 105 and AD 158, the only actual epigraphy on the ground is an altar from Mumrills on the Antonine Wall, which places them there at an unknown date. That altar was set up by *duplicarius* Valerius Nigrinus (*RIB* 2140). The altar raised at Burgh-by-Sands (*RIB* 2045) that is sometimes taken as evidence of this unit cannot be confirmed as such.

Chapter 8

Cohorts

I Afrorum
I Alpinorum
I Aquitanorum
I Asturum
II Asturum
IV Asturum
I Baetasiorum
I Batavorum
III Batavorum
VIII Batavorum
IX Batavorum
I Bracaraugustanorum
III Bracaraugustanorum
IV Breucorum
I Celtiberorum
I Classica
I Cornoviorum
I Cugernorum
I Dacorum
I Delmatarum
II Delmatarum
IV Delmatarum
I Frisiavonum
IV Frisiavonum
II Gallorum
IV Gallorum
V Gallorum
VI Gallorum
I Hamiorum
I Hispanorum
I Lingonum
II Lingonum
III Lingonum
IV Lingonum
I Menapiorum Nautarum
I Morinorum
I Nauticarum
I Nerviorum
II Nerviorum
III Nerviorum
IV Nerviorum
VI Nerviorum
I Pannoniorum
II Pannoniorum
V Pannoniorum
V Raetorum
VI Raetorum
I Sunucorum
I Thracum
II Thracum
VI Thracum
VII Thracum
I Tungrorum
II Tungrorum
I Vangionum
I Vardullorum
II Vasconum

Cohors I Aquitanorum (aka Cohors Primae Aquitanorum Veterana)

Type: Infantry
Size: Quingenary

There is some confusion surrounding this unit's name. A Cohors I Aquitanorum Veterana is attested many times in Germany, and could conceivably be this unit, though it is worth noting that not once in Britain is the epithet 'Veterana' used in an inscription, lending some weight to the idea that they were, in fact, two different units. This is lent further support by the dates of inscriptions, for if they were the same unit, then they must have moved backwards and forwards between the two provinces several times, something not generally found among the *auxilia*. As such, I have set aside German references to the Veterana unit and constructed a history based on the premise that they were two separate units, with the one in Germany taking its epithet to distinguish itself from the younger British unit.

Though I Aquitanorum was clearly raised among the Aquitani, our first evidence of the unit may come from Sardinia. Three tombstones found on the island reference the unit without the 'Veterana' epithet and have been convincingly dated to the reign of Tiberius. It may well be, then, that the unit's first known posting was on the island of Sardinia, perhaps in the role of a policing unit.

Four diplomas place the unit in Britain between the reigns of Hadrian and Antoninus Pius. It seems highly likely that the unit was brought to Britain during Hadrian's shake-up of the provincial military, where it formed the first garrison of the fort at Carrawburgh on his new wall. A building inscription there contains '[...]V[...]O LEG', which has been taken as a reference to the Sextus Julius Severus who was governor circa 130–131. Suggestions that it references instead Gnaeus Julius Verus can be disregarded as the unit was based elsewhere during his tenure, as confirmed by epigraphy. AD 130 is the generally accepted date for the construction of Carrawburgh, and so the inscription would correspond to their part in the initial building work. The unit's presence at Carrawburgh can only have been very brief, perhaps during the fort's construction, circa 130–131, for Cohors I Tungrorum is also attested at Carrawburgh by a building inscription during the Hadrianic era. Perhaps the two units were both drafted in to create the fort (which was an afterthought that was added to the wall two years after completion) and, once it was done, I Aquitanorum was posted elsewhere and I Tungrorum garrisoned their new fort.

Following the brief time at Carrawburgh, the unit seems to have been reassigned to the fort of Navio at Brough-on-Noe. Given the building inscription from there, datable to AD 159 and naming Antoninus Pius, the unit may have been relocated once the Antonine Wall was occupied and Hadrian's Wall became redundant. An altar found at Bakewell (*RIB* 278) cannot be taken as evidence for the unit's presence there, or even the presence of an undiscovered fort, since Bakewell is just 10 miles from Navio, where the unit was definitely based. Neither should lead seals found in Leicester be taken as evidence of the unit's presence there.

The prefect Aemilius Pardalas donated a statue to the House of the Vestals in Rome, to thank the chief vestal, Campia Severina, for her aid in securing his rank and position. This has been dated to AD 240 and gives an interesting insight into the value of patronage in military assignments. The next, and indeed last, location the unit can be associated with is Brancaster, attested by tile stamps. Since this fort was constructed around AD 220, and the known garrison there at the end of Roman Britain was a Dalmatian unit, we can assume that I Aquitanorum formed an earlier, probably even the first, garrison there, given the tile stamp evidence. As such, it is quite possible that the lucky and grateful Aemilius Pardalas was the first commander at Brancaster.

The interesting find of the tombstone of Tarcho Tarbunnis is a matter for open discussion. Though he died in Salona, he may well have served in Britain with this unit, and the word CAM[...] has been taken to suggest he was recruited in Colchester (Camulodunum). Given the likelihood that Tarbunnis, who died in Dalmatia, was part of this unit, then, I have further included Burrius Trebocus, who also references the unit, and who died in Dalmatia.

There is no later mention of the unit, and their fate remains unknown.

Attested officers and men:
Tribune Aemilius Pardalas
Prefect Nepos at Carrawburgh
Prefect Quintus Sittius Caecilianus at Brough-on-Noe
Prefect Proculus at Brough-on-Noe
Prefect Quintus Caecilius Optianus
Prefect Capitonius Priscus at Brough-on-Noe AD 158
Miles Tarcho Tarbunnis
Miles Burrius Trebocus
Rufus Valentinus
Titus Iulius Capito
Decumus Cniensis

Inscription from Brough-on-Noe, now in Buxton Museum. (*Author's collection*)

Inscriptions:
Bitti (Bitta) – tombstone of Decumus Cniensis (*AE* 1920, 96)
Brancaster (Branodunum) – tile stamps (*RIB* 2466)
Brough-on-Noe (Navio) – altar to Mars Braciaca (*RIB* 278)
Brough-on-Noe (Navio) – altar to Hercules Augustus (*RIB* 3181)
Brough-on-Noe (Navio) – building dedication (*RIB* 283)
Carrawburgh (Brocolitia) – building dedication (*RIB* 1550)
Ervace (Aequum) – tombstone of Burrius Trebocus (*CIL* III, 9760)
Ferentino (Ferentinum) – inscription of Caecilius Optanus (*CIL* X, 5831)
Musait-Dunareni (Sacidava) – career inscription of Capitonius Priscus (*AE* 1982, 850)
Oschiri (Castro) – tombstone of Rufus Valentinus (*CIL* X, 7596)
Oschiri (Castro) – tombstone of Titus Iulius Capito (*AE* 1982, 438)
Rome (Roma) – dedication to the chief Vestal (*ILS* 4929)
Solin (Salona) – tombstone of Tarcho Tarbunnis (*CIL* III, 2053)

Cohors II Asturum (aka Cohors Secundae Asturum Equitata Pia Fidelis)

Type: Equitata
Size: Quingenary

Cohors II Asturum clearly had a very active history. Raised from among the Astures in Spain, probably during the reign of Augustus, they seem to have been posted to the quarries at Brohl in Germany, where they are attested in epigraphy. Dating there is difficult, but they were clearly present by AD 80, when they appear on a German diploma.

We know the unit was in Britain by the diploma of AD 105, though the year of their transfer is uncertain. One intriguing possibility is offered by the inscription of Iulius Karus, since, as well as labelling him a prefect of II Asturum and a tribune of Legio III Cyrenaica, it also references a 'BELLO BRITANNICO' (British war) in which he was awarded both the Mural Crown and the Camp Crown, as well as the prized Hasta Pura. The inscription is not easily dated, but the British war referred to cannot reasonably be the invasion of 43, since neither unit he was connected with were present for the invasion. Indeed, since II Asturum were still in Germany in AD 80 and III Cyrenaica were never present on the island, a date prior to AD 80 can be ruled out. An attractive possibility, and one preferred by Birley, is that Karus was assigned to the unit as they were brought from Germany to Britain during the governorship of Agricola, immediately after AD 80, and that he won his awards with them in Scotland and at Mons Graupius. Of course, other troubles cannot be ruled out, including the revolt of AD 119, but the push into Scotland would explain the unit's move to Britain as well as Karus' war.

The unit's first base in Britain seems to have been Llanio, and there is an attractive suggestion that they were brought to the area to oversee the gold mines of Dolaucothi, given their knowledge of quarries and mining. They may, therefore, have also occupied Pumsaint in the area. A move to Caersws in the same era is not impossible, given the find there and its proximity to Llanio. Whatever the case, the unit seem to have been based in Wales prior to the Hadrianic era.

Quintus Baienus Blassianus, as noted by Birley, may have received his commission from A. Platorius Nepos, Governor of Britannia AD 122–124, given that both men's home towns were in the same area, around 20 miles apart. It is quite possible that the unit were involved in the general upheaval during the building of the wall, and they were still present in Britain in AD 127 as noted in a diploma. The interesting thing is that they are also noted in another diploma of AD 127 as being in Germany. This lends heavily to the suggestion that the unit were transferred back to their original province in that year.

The unit may have garrisoned Köln at this point, but their stay in Germany was not to last. By the reign of the Severans, the unit had once more been transferred to Britain, now based on Hadrian's Wall at Great Chesters. This is confirmed by a dedication there mentioning the unit and dated to AD 225. By this time, the cohort has gained the honorifics Severianae Alexandrianae after the emperor. The unit attested at Great Chesters in the *Notitia Dignitatum* is Cohors I Asturum, though there is some debate over this. At first glance it seems that I Asturum is a transcription error in the text, and that it should refer to II Asturum, who are already attested at that site. However, an inscription (*CIL* VIII, 20736) from Souk-el-Ghozlane in Algeria places its subject (Prefect Quintus Gargilius Martialis) as commander of '*cohortis I Asturum provinciae Brittanniae*', which confirms that unit's presence in the province. As such it is at least worth considering the possibility that in the third or fourth centuries, II Asturum were replaced at Great Chesters by their sister cohort, I Asturum, while they themselves moved on to some other unknown garrison not mentioned in the *Notitia*. Of course, given the changes in the military over the fourth and fifth centuries, it is not impossible that the two Asturian cohorts, both seriously reduced, were combined, taking the numeral of the older unit, and it is therefore in effect *both* cohorts that occupy Great Chesters.

The finding of a lead seal at Brough-by-Bainbridge should not be taken as evidence of their occupation there, though it remains a possibility, perhaps between returning from Germany and being assigned to Great Chesters.

Attested officers and men:
Prefect Gaius Iulius Karus, early second century
Prefect Quintus Baienus Blassianus, AD 122
Prefect Gnaeus Munatius Aurelius Bassus, early second century
Prefect Titus Statilius Messalinianus
Prefect Gaius Saturius Secundus
Prefect Publius Furius Rusticus
Imaginifer Gemellus at Brohl first century AD
Miles Titus Flavius Tullio at Köln
Boudus at Caersws

Inscriptions:
Batina (Ad Militare) – tile stamps (*CIL* III, 10674)
Brohl – dedication (*CIL* XIII, 7693)
Brohl – dedication (*CIL* XIII, 7705)
Brough-by-Bainbridge (Virosidium) – lead seal (*RIB* 2411.97)
Caersws – pottery fragment (*AE* 1995, 994b)
Cyrene – career inscription of Iulius Karus (*AE* 1951, 88)
Great Chesters (Aesica) – dedication (*RIB* 1738)

Bronze torc found at Great Chesters, now in the Hancock Museum. (*Author's collection*)

Great Chesters (Aesica) – tile stamps (*RIB* 2467)
Köln (Colonia Agrippensis) – tombstone of Tullio (*AE* 1974, 455)
Lambese (Lambaesis) – tombstone of Furius Rusticus (*CIL* VIII, 18131)
Llanio (Bremia) – building inscription (*RIB* 407)
Llanio (Bremia) – building inscription (*RIB* 408)
Mentana (Nomentum) – inscription of Munatius (*CIL* XIV, 3955)
Pisa (Pisae) – tombstone of Saturius Secundus (*CIL* XI, 1437)
Rome (Roma) – tombstone of Statilius Messalinianus (*CIL* VI, 1850
Trieste (Tergestum) – inscription of Baienus Blassianus (*CIL* V, 539)

Cohors I Baetasiorum (aka Cohors Primae Baetasiorum Civium Romanorum)

Type: Infantry
Size: Quingenary

Raised in Germany among the Baetasii, this unit was likely one of several created in the aftermath of the Civilis Revolt of AD 69–70. It is therefore also a reasonable assumption that the unit accompanied Petillius Cerialis to Britain almost immediately. All we can say for certain of their early movements is that they were in Britain by the diploma of AD 103, and it is entirely possible that they were involved in the Flavian advance into Scotland under Agricola. We are equally unaware of the unit's movements and garrison throughout the Trajanic and Hadrianic periods, though diplomas continue to place them in Britain through to AD 135.

Our first epigraphic record of the unit comes from the Antonine Wall. There has been considerable discussion over their presence there, since they are attested at both Old Kilpatrick and Bar Hill. Some prefer the notion that the unit occupied Old Kilpatrick, but a detachment was stationed alongside the Hamian archers at Bar Hill. Others would suggest two periods of Antonine occupation, with the unit at Old Kilpatrick, and then taking over Bar Hill from the Hamians. We can certainly date the unit's presence at Bar Hill to the period AD 142–165 due to an inscription raised by them there to Antoninus Pius. There is, however, an interesting and enticing possible solution to the question.

The inscription at Old Kilpatrick includes the text COH I BAETASIORVM C R CVI PRAEEST PVBLICIVṢ MATERNVS PRAEF C A IVLIO CANDIDO LEG I ITALICAẸ, which translates as 'The First Cohort of Baetasians, Roman citizens, commanded by the prefect Publicius Maternus, under the charge of Julius Candidus, centurion of the First Legion Italica.'

Though there is no overt way to date this inscription, and an Antonine date is usually assumed on the wall, questions are raised over the inclusion of a centurion of the First Italica. That legion is never attested in Britain, and the closest they ever came was being stationed in Gallia Lugdunensis prior to the events of AD 69. So what was one of their centurions doing in Britain in the Antonine era while his legion was based at Novae in Moesia? The answer might be surprisingly simple. The First Italica was one of the Danubian legions that threw their support behind Septimius Severus in AD 193, and marched on Rome for him. As such, it is not beyond the bounds of reason to suggest a link between this Centurion Julius Candidus and the emperor, and it is possible that Candidus accompanied Severus and his sons on their campaign in Britain in AD 208–211. This would explain why a centurion from a Moesian legion is running I Baetasiorum at Old Kilpatrick, and not their own prefect, who is also mentioned on the altar. If, then, we assign a Severan date to Old Kilpatrick, it is quite possible that during the Antonine period, the only station of the cohort was Bar Hill.

The next place the unit is attested is Maryport, and it seems certain that they were pulled back from Bar Hill when the wall was abandoned in AD 162, then sent to garrison the coastal station. They seem to have spent perhaps two decades or so there, before moving on. Things then become complicated once more. There appear to be three locations linked with the unit in the reign of Severus, all far apart, and all in a short time. A potential sequence can be identified, however.

In AD 208, when Severus arrives in Britain, I Baetasiorum, with decades of experience dealing with coastal watch, are put under the command of one of the emperor's pet centurions. They are assigned to Old Kilpatrick, where their coastal experience will be important, given that fort's value as a supply route for the fleet, and its position, watching over movements among the nearby islands. The unit is not kept there long, and is then withdrawn and sent south. A fragmentary inscription found at Manchester appears to mention the emperor's ill-fated son Geta and the cohort. That the inscription seems to have been smashed into fragments and the metal lettering removed adds weight to this theory. This would place the unit in Manchester between AD 208 and 211. Once again, though, the unit is not kept in garrison there for long. Still in the same campaign they are moved on again, this time to Reculver, on the Kentish coast.

Reculver seems to have been constructed in stone over an earlier fort in around AD 220. Severus died in AD 211, and the Scottish campaign was ended by his son, the imperial family leaving Britain thereafter. It is quite reasonable to suggest that I Baetasiorum, with experience at the coastal stations of Maryport and Old Kilpatrick, were selected for Reculver based on their past. The unit are attested at Reculver with tile stamps, suggesting that the construction of the fort's buildings was still in progress when they arrived. It is highly likely therefore that the building inscription from Reculver's *principia*, which is dated by the governor Triarius Rufinus to AD 210–222, was created by I Baetasiorum. Their stay at Reculver was to be far longer than any other in their previously active history, for the unit is still there at the close of Roman rule, though then commanded by a tribune. The *Notitia Dignitatum* tells of of '*Tribunus cohortis primae Vetasiorum, at Regulbio.*'

Attested officers and men:
Prefect Ulpius Titianus at Maryport
Prefect Titus Attius Tutor at Maryport AD 162–185
Prefect Lucius Tanicius Verus at Bar Hill
Prefect Publicius Maternus AD 208–211

Inscriptions:
Bar Hill – altar to Silvanus (*RIB* 2187)
Bar Hill – altar (*RIB* 2169)
Bar Hill – building dedication (*RIB* 2170)
Manchester (Mamucium) – fragment (*RIB* 581)
Maryport (Alauna Carvetiorum) – altar to Jupiter (*RIB* 830)
Maryport (Alauna Carvetiorum) – altar to Mars (*RIB* 837)
Maryport (Alauna Carvetiorum) – altar to Mars (*RIB* 838)
Maryport (Alauna Carvetiorum) – altar to the emperor's victory (*RIB* 842)
Maryport (Alauna Carvetiorum) – altar to the emperor's victory (*RIB* 843)
Maryport (Alauna Carvetiorum) – altar to Jupiter (BRIT 44.3)

Altar to Jupiter from Old Kilpatrick, now in the Hunterian Museum. (*Author's collection*)

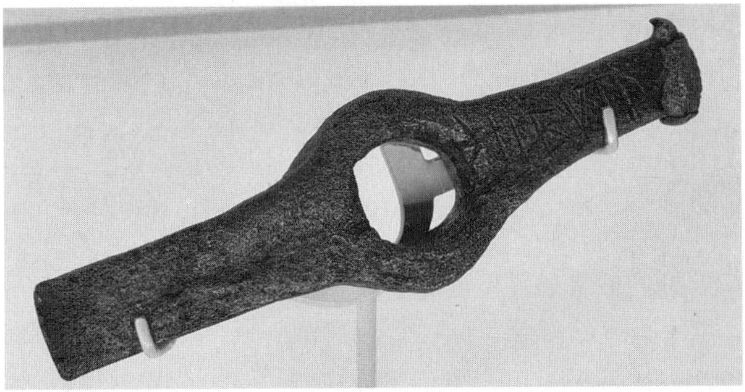

Inscribed Hammer head belonging to Aebutius from Bar Hill, now in the Hunterian Museum. (*Author's collection*)

Old Kilpatrick – Altar to Jupiter (*RIB* 3509)
Reculver (Regulbium) – tile stamps (*RIB* 2468)
Reculver (Regulbium) – construction dedication (*RIB* 3027)

Cohors I Batavorum (aka Cohors Primae Batavorum)

Type: Equitata
Size: Quingenary

The history of much of this unit's service remains unknown. They may have been an early creation, involved in the invasion of AD 43 when Batavians are noted in the Battle of the Medway. They may have been involved in the suppression of the druids at Anglesey in AD 61, when Germanic troops crossed the strait. They may instead have come to Britain with Petillius Cerialis, but whatever the case, they were quite likely to have been one of the Batavian units known to have served at Mons Graupius under Agricola. All of this is perfectly logical and reasonable, though also, sadly, conjectural. Our first direct evidence for them comes from the diploma of AD 122, which places them in Britain by that time. They are only attested for sure in two locations, and possibly a third, and between them, they give us a feasible and straightforward service record.

Two building inscriptions from the Hadrian's Wall fort of Carvoran suggest the unit's presence there, though one omits the 'I' from the unit name. A unit of Hamian archers took up residence at Carvoran around AD 136, and VI Nerviorum seem to have been there in the Hadrianic era, and so it is possible that I Batavorum occupied the fort at some point before that time. Given that it is believed that an earlier fort of Agricolan date lies beneath the extant remains, it is possible that I Batavorum was installed at Carvoran immediately following the Domitianic withdrawal from Scotland, and stayed there until replaced by the Nervians.

Our next evidence is the shakiest part. A dedication from Castlecary on the Antonine Wall names a Batavian cohort, though without a number. If we are to believe that this is indeed our unit, then we can place them on the northern wall in the Antonine era, though this might be wishful thinking, for there is no further evidence there, and Castlecary also records occupation by Cohors I Vardullorum and Cohors I Tungrorum, so while this is a possibility, it should not be relied upon. It may be that a vexillation of a Batavian unit stayed in the fort briefly.

Our next evidence, and the most concrete of all for this unit, comes from the Hadrian's Wall fort of Carrawburgh, where they are attested in no less than eleven inscriptions, including the famous altars from the Mithraeum and

Coventina's Well. It is certain from the epigraphy that the unit was in place there in the Severan era, and they remained the garrison until the removal of Roman government from Britain, for the fifth-century *Notitia Dignitatum* records '*Tribunus cohortis primae Batavorum, at Procolitia.*'

Attested officers and men:
Prefect Aulus Cluentius Habitus at Carrawburgh AD 198–211
Prefect Lucius Antonius Proculus at Carrawburgh AD 212–222
Prefect Marcus Simplicius Simplex at Carrawburgh
Prefect Aelius Tertius at Carrawburgh
Prefect Titus D. [...] Cosconianus at Carrawburgh
Prefect Marcus Flaccinius Marcellus at Carrawburgh
Prefect Marcus Hispanius Modestinus, early third century
Prefect Burrius [...]stus, AD 237
Buccinator Longinus, third century
Centurion [...] Martius

Inscriptions:
Brough (Verteris) – lead seals (*RIB* 2411.99)
Carrawburgh (Brocolitia) – altar to Coventina (*RIB* 1534)
Carrawburgh (Brocolitia) – altar to Covventina (*RIB* 1535)
Carrawburgh (Brocolitia) – altar to Fortuna (*RIB* 1536)

The Mithraeum at Carrawburgh. (*Author's collection*)

Carrawburgh (Brocolitia) – altar to Mithras (*RIB* 1544)
Carrawburgh (Brocolitia) – altar to Mithras (*RIB* 1545)
Carrawburgh (Brocolitia) – altar to Mithras (*RIB* 1546)
Carrawburgh (Brocolitia) – building dedication (*RIB* 1553)
Carrawburgh (Brocolitia) – tombstone of Longinus (*RIB* 1559)
Carrawburgh (Brocolitia) – tombstone of a *signifer* (*RIB* 1560)
Carrawburgh (Brocolitia) – fragmentary tombstone (*RIB* 1562)
Carrawburgh (Brocolitia) – altar to the nymphs and *genius loci* (*RIB* 3316)

Third century tombstone of a signifer from Carrawburgh, now in Chesters Museum. (*Author's collection*)

Carvoran (Magnis) – building inscription (*RIB* 1823)
Carvoran (Magnis) – building inscription (*RIB* 1824)
Castlecary – dedication (*RIB* 2154)
Hadrian's Wall, Milecastle 59 – altar to Mars Cocidius (*RIB* 2015)

Cohors III Bracaraugustanorum (aka Cohors Tertiae Bracaraugustanorum)

Type: Infantry
Size: Quingenary

There is some confusion surrounding the history of this unit. Firstly, there appear to have been at least two units with identical names and numbers, if not three, for one is attested in Raetia, one in Britain, and one in Syria, all during the same eras. There is little chance of all these units being the same one.

Some historians have chosen to identify the Raetian unit as having been transferred to Syria Palaestina and Judea, while others have presumed the British unit was the one moved, and that the unit left Britain circa AD 133 with the governor Sextus Julius Severus when he was summoned by Hadrian to end the Bar Kockhba Revolt.

The problem is that neither unit fits the bill well. If any two of the three attested units are presumed to be the same cohort, then that unit would have needed to be moved backwards and forwards between provinces with dizzying regularity.

The table below records the diploma dates for units called III Bracaraugstanorum in Britain, Syria and Raetia, and show the difficulty in attempting to identify any of them as the same unit.

Britain				103			122	124	127					145				158		
Syria	86	87	90							136		139							160	
Raetia					107	116					138		140		153	156	157			161

Additionally, on most of the Eastern diplomas, the unit is referred to as III Callaecorum Bracaraugustanorum, further suggesting the need to identify this as a completely different unit. As such, despite the time we have devoted to this, it would appear that the cohort with which we are concerned has a very simple history. It was raised in the area of Braga (Bracara Augusta) in Portugal and sent to Britain sometime during, or prior to, Trajan's reign. There, between AD 103 and 127 they seem to have been installed at the fort of Manchester, and are attested there and at nearby Melandra Castle in tile stamps. We simply have no answer to where the unit ended up, as they are not attested after AD 158 and are not mentioned in the *Notitia Dignitatum.*

Stamped tile from Melandra Castle now in Manchester Museum. (*Author's collection*)

Attested officers and men:
No members of the unit have been listed here, since the inscriptions at Rome, Constantine, Pesaro, Arles, Lambese and Segermes could all refer to any unit with this name. As with the unit's history, its membership remains confused.

Inscriptions:
Manchester (Mamucium) – stamped tile (*RIB* 2469, i)
Melandra Castle (Ardotalia) – stamped tile (*RIB* 2469, iv)

Cohors IV Breucorum (aka Cohors Quartae Breucorum Antoninianae)

Type: Equitata
Size: Quingenary

That this unit was a quingenary *equitata* cohort is conjecture, based upon the nature of the known sister units. Raised among the Breuci, an Illyrian tribe, they may have accompanied Legio IX Hispana from their homeland to take part in the invasion of Britain in AD 43. The first dated evidence we have for them in the province is the diploma of AD 122, though we can be reasonably sure they were present earlier, for they are attested in a tile stamp at Castleshaw, which was abandoned following Trajan's reign.

They seem to have garrisoned the fort at Slack, and operated the tilery at nearby Grimescar Wood. Many tiles stamped by the unit have been excavated there, as well as at Castleshaw, Castleford, Manchester and Aldborough. Their time at Slack ended with Hadrian's reorganization of the military, and they are next attested at Bowes in AD 130–133 in a building inscription. That now-lost inscription was originally recorded as 'COH IIII F' and believed to refer to a Cohors IV Frisiavonum. Given the lack of any other evidence for such a unit, the inscription is now assumed to reference a B rather than an F, connecting it instead with IV Breucorum. That the unit's stamped tiles have been found at both other sites connected with the unit, but not at Bowes might suggest that the fort needed no building or repair work during their occupation, or that perhaps their occupation was brief enough not to warrant tile production, or perhaps simply that the tiles exist but have yet to be found. Absence of tiles does lend a little support to the original interpretation of the inscription, which would mean IV Breucorum were not at Bowes after all, though unless the stone is found, this can never be proven either way.

We can comfortably state that the unit was based at Ebchester by AD 213–222 based on an inscription discovered there referencing the unit as Antoninianae. A number of tiles stamped with the unit's name have been found there, an echo of their early posting at Slack. Of numerous finds at Ebchester, only two specifically name the unit and so are included here. Neither the cohort nor the fort are included in the *Notitia Dignitatum*, and so we have no record of their service after the Severan era. A bronze trulla (*AE* 1995, 1002) now in the possession of Louth museum references the unit, along with the inscription 'CL SENIORIS ANNI', though its origin is unknown, and cannot be connected to any site with certainty.

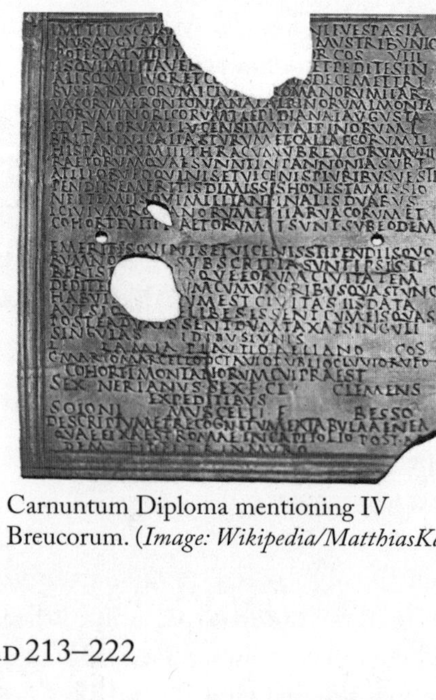

Carnuntum Diploma mentioning IV Breucorum. (*Image: Wikipedia/MatthiasKabel*)

Attested officers and men:

Actarius Iulius Gr[...]nus at Ebchester AD 213–222

Claudius Senioris Anni

Inscriptions:
Bowes (Lavatris) – inscription to Hadrian (*RIB* 739)
Ebchester (Vindomora) – dedication to Minerva (*RIB* 1101)
Ebchester (Vindomora) – building inscription (*RIB* 1113)
Slack (Cambodunum) – tile stamps (*RIB* 2470.1–3)

Cohors I Classica (aka Cohors Primae Aelia Classica Pia Fidelis)

Type: Infantry
Size: Quingenary

This unit may have begun life as the Cohors Nautarum, drawn from marines of the fleet in Frejus, a unit that is attested in the area of Nice/Cimiez. They may well then have changed their name to Cohors I Classica, which is attested in Germany on a diploma of AD 80, before moving to Britain and acquiring their honorific Aelia.

Though it is conjectured that they arrived with Hadrian and were involved in the building of the wall, gaining their honorific then, the only evidence for this is one inscription from the wall, which can be readily discounted. I have not included the building inscription *RIB* 3297, which says CH I CL AVGV, as this is less likely to reference a member of Cohors I Classica than the First Cohort under Claudius Augu... Their presence in Britain is attested in diplomas from AD 145–158, and so it seems more likely that they acquired 'Aelia' from Antoninus Pius. For the possibility of appearing in an earlier diploma under another name, see Cohors I Nauticarum.

In AD 158, it appears that they once more returned to Germany, for they are attested there on another diploma of that year. We have no record of their next return to Britain, though it certainly occurred, perhaps during the campaigns of Septimius Severus. With the exception of the lead seals found at Ravenglass, which cannot be taken as evidence of occupation, only one other piece of evidence on the island references them. The fifth-century *Notitia Dignitatum* tells us of a '*Tribunus cohortis primae Aeliae classicae, at Tunnocelo.*' Tunnocelum is an as-yet unidentified fort, though it is reasonable to assume a coastal location in Cumbria, since it is preceded in the *Notitia* by Gabrosentum (the fort at Moresby). This is the last we hear of them.

Attested officers and men:
Prefect Lucius Volusius Maecianus at Ostia, Hadrianic
Prefect Claudius Paternus Clementianus at Epfach
Prefect Quintus Livius Velanius Pius Severus

Tubicen Apolonius Dionysus at Cimella
Duplicarius Titus Iulius Rem[...] at Nice
Miles Lucius Valerius Verecundus Rutaenus at Köln
Miles Lucius Nonius Rufus at Anzio
Sextus Nonius Severus
Gaius Lucius Ani[...] at Frejus
Nonius Quadratus at Nice
Sextus Valerius Maxsumus at Eysses
Valerius Gallus at Eysses

Inscriptions:
Anzio (Antium) – tombstone of Nonius Rufus (*CIL* X, 6672)
Cimella – tombstone of Apolonius (*CIL* V, 7884)

Sailors depicted on Trajan's Column. (*Author's collection*)

De Meern – Tile Stamps (*AE* 192, 196)
Epfach (Abudiacum) – career inscription (*AE* 1968, 406)
Eysses – tombstone of Sextus Valerius (*CIL* XIII, 923)
Eysses – tombstone of Valerius Gallus (*CIL* XIII, 924)
Frejus (Forum Iulii) – tombstone of Gaius Lucius (*AE* 1904, 7)
Köln (Colonia Agrippensis) – tombstone of Valerius Verecundus (*AE* 1906, 153)
Monteleone – career inscriptions of Livius Velanius (*CIL* IX, 4885 and 4886)
Nice – tombstone of the Nonii (*CIL* V, 7892)
Nice – tombstone of Titus Iulius (*CIL* V, 7887)
Ostia Antica (Ostia) – career inscription (*AE* 1955, 179)
Ravenglass – lead seals (*RIB* 2411.94)

Cohors I Cugernorum (aka Cohors Primae Ulpia Traiana Cugernorum Civium Romanorum)

Type: Infantry
Size: Quingenary

This unit was raised from among the Cugerni near Xanten in Germany and, interestingly, just like the unit, that town was given the name Ulpia Traiana by Trajan. A cohort being honoured with two of the emperor's names is very unusual, and it is likely that they began as Cohors I Ulpia Cugernorum, raised there by Trajan, perhaps at the very close of the first century, specifically to bolster his army for the coming invasion of Dacia. Trajan had strong links with Germany, having served there as a military commander and governor, and so the inclusion of German troops would be only natural. It is possible, then, that the unit's first action was seen in Dacia, raised and sent there for the First Dacian War of AD 101–102, and it is possible that the unit received its second imperial epithet then. There is an argument against this, in that the first British diploma in AD 103 names them only I Cugernorum, though it seems likely they would have at least one honorific by then, and we may simply be looking at omission.

In AD 103, then, we know the unit was in Britain, perhaps settled there after the war. It seems unlikely they would have been involved in the Second Dacian War of AD 105–106, having been moved to Britain by then, at the opposite end of the empire. Their presence in the province is confirmed in diplomas of AD 103, 122, 124 and 127, though we cannot be certain of their movements in the Trajanic and Hadrianic eras. It is possible that the altar of Carrawburgh hints at a presence there, but it cannot be taken as a certainty, for it is highly likely that officers from other garrisons in the region voted altars at this important shrine.

Our first datable evidence comes from the area of Cramond in Scotland, where a milestone of the reign of Antoninus Pius, marking the road from Cramond to Trimontium, names this unit. The milestone stood perhaps 4 miles from the Cramond fort, and the unit was therefore likely based there during the Antonine era. It seems that Cohors II Tungrorum were the first Antonine garrison there, so perhaps I Cugernorum formed the garrison in the second Antonine phase. The fort is believed to have been abandoned around AD 170, and so by that date the unit had moved on, though where it then went we cannot be sure. The recent find of a ring-pommel sword pendant at Cramond, dated to the Severan era, raises questions. Such items are identified with Germanic troops, and may point to a second period of occupation by I Cugernorum during the Severan campaigns. However, this evidence is at best circumstantial, and does not answer where the unit was from AD 170.

Milestone from Ingliston House, near Cramond. (*Author's collection*)

Following the withdrawal from Scotland in AD 211 under Caracalla, the unit is next attested at Newcastle-upon-Tyne in a dedicatory inscription dated to AD 213. It is quite possible that the altar at Carrawburgh dates to this period of occupation, an officer from Newcastle travelling some 25–30 miles along the wall to the popular shrine.

The finds at Osijek and Konuralp probably both reference retired former commanders of the unit and certainly cannot be taken as evidence of the unit's presence. Indeed, our evidence for the unit ends with the Severan era. Only one possibility remains to be mentioned. The fifth-century *Notitia Dignitatum* tells us of the garrison of Newcastle at the close of Roman Britain: '*Tribunus cohortis primae Cornoviorum, at Ponte Aeli.*' This seems to reference an auxiliary unit drawn from the Cornovii in central England. It is suspicious, though, that there is no other reference to this unit, and it breaks the standard of not basing a unit raised in a province *within* that province. The unlikeliness of this cohort leads me to wonder whether it represents a transcription error in the *Notitia*, and that the author mistook Cugernorum for Cornoviorum, which would then suggest that I Cugernorum remained the garrison of Pons Aelius until the end.

Attested officers and men:
Prefect Lucius Domitius Proculus at Konuralp, second century
Prefect Pontius Antoninus at Osijek
Aurelius Campester at Carrawburgh

Inscriptions:
Carrawburgh (Brocolitia) – altar to Coventina (*RIB* 1524)
Cramond – milestone of Antoninus Pius (*RIB* 2313, edited *AE* 1975, 581)
Konuralp (Prusias ad Hypium) – dedicatory stone (*ILS* 8866)
Newcastle-upon-Tyne (Pons Aelius) – dedication to Julia Domna (*RIB* 3284)
Osijek (Mursa) – career inscription (*AE* 1974, 535)

Cohors I Dacorum (aka Cohors Primae Aelia Dacorum Augusta Antoniniana Maximiniana Gordiana Postumiana, Tetricianorum Probiana Milliaria)

Type: Infantry
Size: Milliary

Following Trajan's conquest of Dacia and its incorporation into the empire, that emperor recruited heavily from the Dacian peoples. Upon Trajan's passing, though Hadrian let some of the conquered territory go, he also consolidated the rest and continued to recruit, as did Antoninus Pius thereafter. Given the unit's Aelia honorific and its presence in Britain on the diploma of AD 126, it seems almost certain that it was Hadrian who recruited this unit.

The first evidence we have of the unit in Britain, and indeed, *all* such evidence, comes from Birdoswald. The fort there was originally of turf and timber, though a new stone fort was constructed in AD 130, and it may be that I Dacorum was the unit brought in to occupy that new fort. Though there is some indication (by way of one building inscription) that Cohors I Thracum worked on the fort in the Severan era, alongside I Dacorum, this should not be taken as evidence of occupation by that unit. They also worked on the wall near Benwell, and were probably brought in simply to aid construction.

Cohors I Dacorum seems to have occupied the fort from some time after AD 126 until the end of Roman government, for they are still placed there in the fifth-century *Notitia Dignitatum*. On the surface, then, their career seems fairly straightforward. Raised in Dacia, transferred to Britain, installed at Birdoswald, and there they stayed until the end.

Despite lone inscriptions at Lorch, Lambese, and Cesena, these should not be taken as a sign of occupation, but of individual retirees. The altar found at

Bewcastle that references the unit should not necessarily point to occupancy, either. Bewcastle is just 6 miles from Birdoswald, and the donation of a visiting officer is a very reasonable solution, though we cannot rule out an early Hadrianic garrison at Bewcastle before a move to Birdoswald. For the possibility of a connected tombstone at High Rochester, see Cohors I Delmatarum.

The one thing that is perhaps of interest about the unit beyond their career path is their name… that and that one of their tribunes had been a member of the First Praetorian Cohort. It would appear that I Dacorum were determined to be survivors. Despite the fact that their service largely covered the troublesome third and fourth centuries, when emperors came and went with wild abandon (*literally* sometimes, and often at the point of a blade), I Dacorum seem to have swung to and fro with the prevailing wind throughout their service. They have been known in their career by the honorifics Aelia, Augusta, Antoniniana, Maximiniana, Gordiana, Postumiana, Tetricianorum, and Probiana. Quite simply whoever now sat on the throne holding the bloody knife, the I Dacorum were immediately and very vocally his supporters. They even took the names of the breakaway emperors of the independent Gallic Empire before then throwing their support behind Probus once back with Rome. That they managed three centuries in Britain almost entirely in the same garrison suggests that their tactic worked well.

Attested officers and men:
Tribune Titus Claudius Proculus Cornelianus at Lambese
Tribune Terentius Valerianus at Birdoswald
Tribune Ammonius Victorinus at Birdoswald
Tribune Statius Longinus at Birdoswald
Tribune Iulius Saturninus at Birdoswald
Tribune Domitius Honoratus at Birdoswald
Tribune Reginius Iustinus at Birdoswald
Tribune Funisulanus Vettonianus at Birdoswald, third century
Tribune Aurelius Iulianus at Birdoswald AD 205–208
Tribune Marcus Claudius Menander AD 219
Tribune Flavius Maximianus at Birdoswald AD 235–238
Tribune Aurelius Faustus at Birdoswald AD 237
Tribune Marcius Gallicus at Birdoswald AD 259–268
Tribune Probius Augendus at Birdoswald AD 259–268
Tribune Pomponius Desideratus at Birdoswald AD 270–273
Tribune Aurelius Verinus at Birdoswald AD 276–282
Tribune Aurelius […] at Birdoswald
Prefect Aulus Fabius Proculus at Tusculum

Acting commander Aurelius Saturninus at Birdoswald
Acting commander Centurion Iulius Marcellinus at Birdoswald
Commander Flavius [...] at Birdoswald
Centurion Decius Saxa
Beneficarius [...]rinus at Birdoswald
Decibalus
Septimus

Inscriptions:
Bewcastle (Fanum Cocidi) – altar to Jupiter (*RIB* 991)
Birdoswald (Banna) – altar to Cocidius (*RIB* 1872)
Birdoswald (Banna) – altar to Jupiter (*RIB* 1874)
Birdoswald (Banna) – altar to Jupiter (*RIB* 1875)
Birdoswald (Banna) – altar to Jupiter (*RIB* 1876)
Birdoswald (Banna) – altar to Jupiter (*RIB* 1877)
Birdoswald (Banna) – altar to Jupiter (*RIB* 1878)
Birdoswald (Banna) – altar to Jupiter (*RIB* 1879)
Birdoswald (Banna) – altar to Jupiter (*RIB* 1880)
Birdoswald (Banna) – altar to Jupiter (*RIB* 1881)
Birdoswald (Banna) – altar to Jupiter (*RIB* 1882)
Birdoswald (Banna) – altar to Jupiter (*RIB* 1883)
Birdoswald (Banna) – altar to Jupiter (*RIB* 1884)
Birdoswald (Banna) – altar to Cocidius and Jupiter (*RIB* 1885)
Birdoswald (Banna) – altar to Jupiter (*RIB* 1886)
Birdoswald (Banna) – altar to Jupiter (*RIB* 1887)
Birdoswald (Banna) – altar to Jupiter (*RIB* 1889)
Birdoswald (Banna) – altar to Jupiter (*RIB* 1890)
Birdoswald (Banna) – altar to Jupiter (*RIB* 1891)

Dedication slab from Birdoswald, now in the Hancock Museum. (*Author's collection*)

Birdoswald (Banna) – altar to Jupiter (*RIB* 1892)
Birdoswald (Banna) – altar to Jupiter (*RIB* 1893)
Birdoswald (Banna) – altar to Jupiter (*RIB* 1894)
Birdoswald (Banna) – altar to Jupiter (*RIB* 1896)
Birdoswald (Banna) – altar to Mars (*RIB* 1898)
Birdoswald (Banna) – altar to Jupiter (*RIB* 1906)
Birdoswald (Banna) – Severan dedication (*RIB* 1909)
Birdoswald (Banna) – dedication (*RIB* 1914)
Birdoswald (Banna) – tombstone of Decibalus (*RIB* 1920)
Birdoswald (Banna) – tombstone of Septimus (*RIB* 1921)
Birdoswald (Banna) – fragmentary dedication (*RIB* 1922)
Birdoswald (Banna) – altar to Jupiter (*RIB* 3438)
Birdoswald (Banna) – altar to Neptune (*BRIT.* V 462)

Silver dedication to Cocidius from Bewcastle, possibly connected to I Dacorum, now in Tullie House Museum. (*Author's collection*)

Birdoswald (Banna) – building stone of Decius Saxa (*RIB* 1918)
Cesena (Caesena) – mutilated inscription (*CIL* XI, 558)
Hadrian's Wall, Denton – building inscription (*RIB* 1365)
Lambese (Lambaesis) – inscription of Cl. Proculus Cornelianus (*AE* 1956, 123)
Lorch – fragment (*AE* 1957, 113)
Tuscolo (Tusculum) – tombstone of Fabius Proculus (*CIL* XIV, 2618)

Cohors I Delmatarum (aka Cohors Primae Delmatarum)

Type: Infantry
Size: Quingenary

It is worth noting that the name of this cohort, and other such units, drawn from the people of the Dalmatian coast in modern Croatia, is spelled in different ways. A and E are often interchangeable, making Delmatarum and Dalmatarum both possible.

The unit is first attested in Britain in a diploma of AD 122, and is still present in one of AD 135. We have no epigraphic record in Britain that predates the Antonine era, though the inscription of Nasellis in Benevento has been given a Trajanic or Hadrianic date, and that of Victor in Saintes a post-Claudian first-century date. It is thus likely that the unit was operating either in Britain or elsewhere from at least the late first century. It is possible that the

Saintes inscription references another unit, though I Delmatarum has been widely accepted.

Our main source of epigraphy is Maryport, where I Delmatarum are given a confirmed date between AD 138 and 165, since the inscriptions reference the emperor Antoninus Pius, and the unit clearly took control there on the departure of I Hispanorum in the Antonine era. Based on the number of Jupiter altars there, it can be assumed that their stay at Maryport consisted of a minimum of four years and likely considerably longer.

They are attested on a building inscription at Chesters, which has been given an Antonine date on stylistic grounds. If the Hadrianic garrison of Chesters was possibly Ala Augusta Ob Virtutem, as is suggested by other finds, there are two possible explanations for the Delmatarum inscription. Either they remained based at Maryport, but spent some time on building work at Chesters, while the focus was on moving the frontier north, or more likely they left Maryport around the end of the Antonine era, replaced by Cohors I Baetasiorum, and occupied Chesters from then until the Asturian *ala* arrived there around AD 175.

A tombstone at High Rochester records the deceased's unit as I DA[...] Whether this references I Dalmatarum or I Dacorum is not known, but alone it cannot be taken as proof of occupation by either, and may represent a lone centurion away from his garrison. I have included him here anyway, given that High Rochester, at a distance of less than 20 miles, is closer to the potential base of I Delmatarum in Chesters than it is to that of I Dacorum at Birdoswald. I have left out of my listings the tombstone from Tuscolo that mentions only 'I DA[...]', as well as that from Balaca in Pannonia that confirms a 'DALMA' unit, but without a number, neither region having confirmed links with this unit. It is worth noting that the Aquino dedication to Ceres by a tribune Iuvenalis of I Delmatarum suggests a close family link to the poet Juvenal, who also hailed from Aquino.

Though we have no evidence of the unit in Britain after the mid- to late second century, one intriguing possibility arises. The fifth-century *Notitia Dignitatum* mentions units garrisoning both the Saxon Shore fort of Brancaster, and the unidentified fort of Praesidio (possibly Bridlington), as 'Equitum Dalmatarum', Since there is no other evidence of a Dalmatian cavalry unit at any time in Britain, it is tempting to see I Delmatarum as the origin of these late Roman coastal cavalry forces.

Attested officers and men:
Tribune [...]nius Iuvenalis at Aquino
Tribune Gaius Iulius Victor at Saintes
Prefect Postumius Acilianus at Maryport AD 138–165

Dedication to Jupiter from Maryport, now in the Senhouse Museum. (*Author's collection*)

Prefect Lucius Caecilius Vegetus at Maryport
Prefect Marcus Nasellius Sabinus at Benevento
Centurion Aurelius Ex[…]imi at High Rochester

Inscriptions:
Aquino (Aquinum) – inscription to Ceres (*ILS* 2926)
Benevento (Beneventum) – building inscription (*CIL* IX, 1618)
Chesters (Cilurnum) – building inscription (*RIB* 3300)
High Rochester (Bremenium) – tombstone of Aurelius (*RIB* 1289)
Maryport (Alauna Carvetiorum) – altar to unnamed deities (*RIB* 810)
Maryport (Alauna Carvetiorum) – altar to Jupiter (*RIB* 831)
Maryport (Alauna Carvetiorum) – dedication to Jupiter (*RIB* 832)
Maryport (Alauna Carvetiorum) – altar to Jupiter (*RIB* 833)
Maryport (Alauna Carvetiorum) – fragmentary inscription (*RIB* 847)
Maryport (Alauna Carvetiorum) – dedication slab (*RIB* 850)
Saintes (Mediolanum Santonum) – dedicatory frieze (*CIL* XIII, 1042–1045)

Cohors IV Delmatarum (aka Cohors Quartae Delmatarum)

Type: Infantry
Size: Quingenary

This unit was raised in Dalmatia, probably in the early first century. Their first posting seems to have been at Bingen in Germany, where three of the unit's tombstones have been found, all dated to the first century. There is a small possibility that the unit served in Trajan's First Dacian War of AD 101–102, given that the tombstone of Antonius Super at Sarmizegetusa in Romania was recorded by Mommsen as naming him tribune of Coh IV Delmatarum, although this is now considered to be erroneous, and that the inscription actually names III Delmatarum.

Either way, the cohort was in Britain by AD 103, as recorded in a diploma. It is possible that their first garrison was at Castle Collen in Wales. A fragmentary inscription there is far from certain evidence, but has been tentatively expanded to refer to both Trajan and a Delmatarum cohort. J.L. Davies puts forward a convincing argument that IV Delmatarum is the most likely option of the three cohorts bearing the name, and thus we can tentatively suggest Trajanic occupation of Castle Collen by the unit.

What we can say with some comfort is that IV Delmatarum was based at the impressive fort of Hardknott under Hadrian when they dedicated a building to him between AD 117 and 127. Hardknott was a relatively short-lived fort, built under Hadrian (perhaps even entirely by this cohort) and abandoned by AD 140. It was reoccupied around AD 160 and occupied once more for only about forty years before being abandoned once more. The reason for the temporary absence of a garrison might coincide with the Antonine push into Scotland. As such, and given that IV Delmatarum are the only unit attested at Hardknott, it might be that the cohort left the fort, campaigned in Scotland, helped build and occupy the northern wall, and then returned to Hardknott when the frontier was pulled back once more. An alternative possibility is that another unit took control of Hardknott in its second period, and that the Castle Collen inscription actually

The fort of Hardknott, Cumbria. (*Author's collection*)

refers to Antoninus Pius, when rebuilding work there is known, suggesting that IV Delmatarum was there *after* Hardknott, and not before. The unit's later garrisons and fate are unrecorded.

Attested officers and men:
Tribune Publius Antonius Super
Prefect Titus Iunius Galerius Severus at Denia, second century
Prefect Pactumerius at M´Daourouch
Prefect Lucius Aprius Claudius
Miles Annius Daverzeus at Bingen
Miles Bato at Bingen
Miles Beusas at Bingen

Inscriptions:
Bingen (Bingium) – tombstone of Annius (*CIL* XIII, 7507)
Bingen (Bingium) – tombstone of Bato (*CIL* XIII, 7508)
Bingen (Bingium) – tombstone of Beusas (*CIL* XIII, 7509)
Bribir (Varvaria) – inscription of Aprius (ILJug 2, 825)
Castle Collen – fragmentary inscription (*RIB* 414)
Denia (Dianium) – inscription of Galerius Severus (*CIL* II, 3583)
Hardknott (Mediobogdum) – dedication to Hadrian (*RIB* 3219)
M´Daourouch (Madauros) – inscription of Pactumerius (*CIL* VIII, 16870)

Cohors I Frisiavonum (aka Cohors Primae Frisiavonum)

Type: Infantry
Size: Quingenary

The origin of this unit is unknown, though they were drawn from the Frisians around the mouth of the Rhine, probably following the Civilis Revolt, circa AD 70. They are attested in a diploma of AD 105, but their movements and garrisons throughout the first and second century remain a mystery. They are attested at both Manchester and Melandra Castle with building inscriptions and centurial stones, and given the proximity of the two sites, it seems certain that they were based at one site during both forts' periods of work. We are therefore handed something of a quandary.

Melandra Castle is believed to have been rebuilt in stone either in the last days of Trajan's reign, or the early days of Hadrian's. The previous fort was of turf and timber, and the site was abandoned in the Antonine period, circa AD 140. Thus we can only see I Frisiavonum at work there around AD 110–120. Manchester was a turf and timber Agricolan fort, and was abandoned circa AD 140, rebuilt in turf and timber again around AD 160, and only built in stone from around AD 198 in the Severan era. With I Frisiavonum recorded on centurial stones at both sites, it seems they were at work at Melandra Castle in circa AD 110, and in Manchester in the Severan era.

Since we know from inscriptions that they were based elsewhere by the mid-third century, the only answer that leaps to mind is two distinct garrisons, despite the fact that the two forts are only 12 miles apart. The suggestion of the above timings is that the unit was the first occupant of Melandra Castle, from the Agricolan era, perhaps even building the initial timber fort. Around the end of Trajan's reign, they rebuilt the fort in stone. We are then left with an uncomfortable break. They must have left Melandra Castle around AD 140, but at that time Manchester had also been abandoned. So sometime after circa AD 120 they seem to have moved to an unknown location before their arrival at Manchester in 160 or thereafter. We cannot even guess at such a move, though the upheavals of the Hadrianic and Antonine eras as units were moved north for the work on two walls mean that many units' movements will probably never be adequately explained. They may have stayed at Melandra Castle until its abandonment around AD 140 and then been moved up to Scotland with the new emperor's northern expansion.

Whatever they were up to in those years, they must have been back in Manchester later in the century. They were perhaps the unit who rebuilt the timber fort there around AD 160, following the abandonment of the Antonine Wall, or perhaps they remained in the north for a little longer and returned to Manchester only at the start of the Severan era. The fort was rebuilt in stone from AD 198, and from the centurial inscriptions I Frisiavonum were involved.

They cannot have garrisoned Manchester for long after its stone rebuild, for they are attested by the mid-third century at the latest at the wall fort of Rudchester, where three altars, all dated to the third century, testify to their presence. Interestingly, pottery manufactured at Corbridge, just 9 miles from their base at Rudchester, may represent the retirement career of a soldier of this unit. Centurion Cudrenus is noted on *RIB* 579 as one of the builders in stone of Manchester in the closing days of the second century. Pottery from Corbridge, including a mortarium, dated to the late second century, bear the maker's stamp 'CVDRE'. It may be, therefore, that Cudrenus was involved in building Manchester, moved to Rudchester with his cohort immediately thereafter, and then retired to nearby Corbridge and opened what appears to have been a thriving pottery business.

The one altar found at Carrawburgh that names the unit should not be taken as evidence of occupation, since the two forts are less than 20 miles apart, and it

Reconstructed fort gate at Manchester. (*Author's collection*)

is quite reasonable that an officer at Rudchester might commission an altar at such an important cult site as Coventina's Well. Their move to Rudchester appears to have been their last, for they are still noted occupying that site at the end of Roman Britain in the *Notitia Dignitatum*.

Dedication to Sol Invictus from Rudchester, now in the Hancock Museum. (*Author's collection*)

Attested officers and men:
Prefect Publius Aelius Titullus at Rudchester, third century
Prefect Tiberius Claudius Decimus Cornelius Antonius at Rudchester, third century
Centurion Masavo at Manchester
Centurion Quintianus at Manchester
Centurion Cudrenus at Manchester
Centurion Candidus at Manchester
Centurion Valerius Vitalis at Melandra Castle
Optio Mausaeus at Carrawburgh
Aponius Rogatianus at Rudchester, third century

Inscriptions:
Carrawburgh (Brocolitia) – altar to Coventina (*RIB* 1523)
Manchester (Mamucium) – centurial stone (*RIB* 577)
Manchester (Mamucium) – centurial stone (*RIB* 578)
Manchester (Mamucium) – centurial stone (*RIB* 579)
Manchester (Mamucium) – centurial stone (*RIB* 580)
Melandra Castle (Ardotalia) – centurial stone (*RIB* 279)
Rudchester (Vindobala) – altar to Mithras (*RIB* 1395)
Rudchester (Vindobala) – altar to Sol Invictus (*RIB* 1396)
Rudchester (Vindobala) – altar to Sol Apollo Mithras (*RIB* 1397)

Cohors II Gallorum (aka Cohors II Gallorum Severinae Alexandrinae Veterana Equitata)

Type: Equitata
Size: Quingenary

Raised among the tribes of central and northern Gaul, the first record we have of the cohort is a diploma of AD 99 which places them in Moesia. A second diploma confirms their presence there in AD 105.

We then have an interesting aside. Before moving to Britain, the evidence initially suggests the unit moved to Mauretania Caesariensis, for they are named on a diploma there in AD 107. There is a notion that units were transferred there to recruit (see I Nerviorum) and this is possibly the case. If II Gallorum had been involved in Trajan's Dacian Wars (for they had been based in the region), perhaps they were seriously under strength by AD 105 and needed a recruiting drive. However, every commander attested in Britain was a prefect, while on the inscription of Gaius Bellicius Primus from Verona, he is named as the tribune of Cohors II Gallorum. While it is not unknown for a tribune to command such a unit, it is not common, and may suggest that a second unit by the name, and one that warranted a tribune in command, was stationed in Mauretania while our unit was in Moesia and Britain. This is, of course, conjecture, but it does neatly explain the two issues at once. Indeed, if there was reason for any confusion between two identically named units, that might also explain the appearance by AD 178 of the epithet 'Veterana' in the unit's name.

What we can say with certainty from the diplomas is that the unit was in Britain from AD 122. The only confirmed evidence we have of their garrison activity comes from the fort of Old Penrith. Since II Gallorum is the sole unit attested at Old Penrith, it is entirely possible that they were its only garrison from AD 122 until its abandonment, replacing some unknown unit when they were brought over by Hadrian, though there may be a case for placing them at Kirkby Thore, (see below).

The unit may have been the recipient of a large influx of German recruits during their time at Old Penrith, for *RIB* 920 refers to a 'Vexillatio Germanorum [Voredensium?]' while *RIB* 919 names a 'Vexillatio Marsacorum', the Marsacii also being a German tribe.

One unique thing about II Gallorum is that though Lucius Naevius Verus Roscianus is mentioned on an altar at Piacenza, it would appear the altar was actually raised in Britannia, and taken home with Verus upon his retirement from the cohort.

Why the unit received the honorific in the name of Severus Alexander we cannot conjecture.

The eventual fate of the unit is unknown. They are last attested AD 222–235, are not mentioned in the *Notitia Dignitatum* and Old Penrith seems to have been abandoned by the early fourth century. There is a notion that they were transferred away because of their Gallic leanings after the fall of the breakaway Gallic Empire and its recapture by Rome in AD 274, though there appears to be no direct evidence of this. One intriguing possibility remains. *RIB* 782, from either Brougham or Kirkby Thore, has been interpreted as referring to a 'Gallorum' unit, although the text is unclear and the stone now lost for study. Alone, this would be poor evidence, but *RIB* 765 from Kirkby Thore is an altar raised to the good luck of the cavalrymen, and the only unit associated with it is I Hamiorum, an infantry unit. As such, it is possible that in the third century, following their time at Old Penrith, II Gallorum were pulled back south to Kirkby Thore.

Attested officers and men:
Tribune Gaius Bellicius Primus at Verona
Prefect Titus Visulanius Crescens at Bologna
Prefect Aurelius Attianus at Old Penrith
Prefect Titus Domitius Hieron at Old Penrith AD 178
Prefect Lucius Naevius Verus Roscianus at Piacenza
Prefect Iulius Lupus at Old Penrith
Prefect Aurelius [...] at Old Penrith, AD 222–235
Prefect Valerius Laetus at Otricoli
Actor Iulius Augustalis at Old Penrith
Trooper Dacus in Britannia (Diploma RMD 184)
Miles Marcus Antonius Rufus in Moesia (Diploma *CIL* XVI, 44)

The fort's east gate at Old Penrith. (*Author's collection*)

Inscriptions:
Bologna (Bononia) – inscription of Crescens (*CIL* XI, 709)
Old Penrith (Voreda) – altar to Jupiter (*RIB* 915)
Old Penrith (Voreda) – altar to Jupiter (*RIB* 916)
Old Penrith (Voreda) – altar to Jupiter (*RIB* 917)
Old Penrith (Voreda) – altar to Mars Belatucadrus (*RIB* 918)
Old Penrith (Voreda) – altar to the Mother Goddesses (*RIB* 919)
Old Penrith (Voreda) – altar to the Mother Goddesses (*RIB* 920)
Old Penrith (Voreda) – dedication (*RIB* 929)
Otricoli (Ocriculum) – inscription of Laetus (*CIL* IX, 4784)
Piacenza (Placentia) – altar to Minerva (*CIL* XI, 1303)
Verona (Verona) – inscription of Bellicius Primus (*AE* 1991, 821)

Cohors IV Gallorum (aka Cohors Quarta Gallorum Equitata Severinae Alexandrianae Probiana)

Type: Equitata
Size: Quingenary

The earliest record of this unit, raised among the Gaulish tribes, is uncertain. A cohort bearing this name that was in service in Hispania under Tiberius may well be this one, or perhaps another unit. Our first concrete, or rather bronze, evidence of the unit, is a diploma of AD 75 that places IV Gallorum in Moesia. Diplomas confirm their presence there until at least AD 105, where they appear to have been based at the fort of Sacidava and worked on the road network, for they raised a milestone naming Trajan, which has been dated rather precisely to

10 December AD 103. That the unit took part in one or both of Trajan's Dacian Wars seems inevitable.

Following the wars, the unit was moved south to Thracia, as attested on a diploma of AD 114, though the diploma of AD 109 from Mauretania names a Cohors IIII Gallorum Civium Romanum and, if this is the same unit and not simply another with a similar name, this may denote a brief stay in Mauretania Tingitana between AD 105 and AD 114, perhaps as a brief recruiting drive.

Their stay in Thracia cannot have been long. Our first diploma mentioning the unit in Britain dates to AD 122, but they must have been present before this. They are attested in the fort of Templebrough quite solidly, with a number of stamped tiles and two tombstones. Templebrough, like so many forts, displays a series of builds on the site, with the earliest dating from the initial Roman campaigns into the north. The second iteration seems to have been a Trajanic construction, and it is in this layer, below the later rebuilds, that the tiles were found. Since we know the unit was in Thracia in AD 114, and Trajan died in AD 117, we have a very narrow window for their transfer to Britain. It would seem that sometime in those three years they arrived on the island, were sent to Templebrough, and reconstructed the fort there. We have no record of their activity in the Hadrianic era, and it is entirely possible that they remained in Templebrough throughout the period. The two tombstones found there have both been dated to the second century, adding weight to this theory.

With the Antonine advance into Scotland and the creation of the Antonine Wall, IV Gallorum was installed as the first garrison of Castlehill fort, circa AD 144–160. The altar raised there to the goddesses of the parade ground appears to have been deliberately buried, probably under said parade ground, upon the withdrawal from the wall.

Two inscriptions from Risingham suggest that when the forces were pulled back to Hadrian's Wall, around AD 160, IV Gallorum were installed in Risingham, one of the wall's outpost forts. The altar found there references 'emperors' in the plural, which heavily suggests either the joint reigns of Marcus Aurelius and Lucius Verus (ending in AD 169 with Verus' death), or of Marcus Aurelius and Commodus (beginning in AD 177). The building inscription found at High Rochester explicitly mentions only a vexillation of IV Gallorum, and so it seems clear that the unit remained in Risingham, but sent a workforce less than 10 miles up the road to High Rochester.

The unit is attested at Castlesteads on the wall by two altars. By AD 213, we can confirm that the unit was in garrison at Chesterholm, where they are well attested in the third century. Indeed, they acquired honorific titles under both Severus Alexander and Probus during their stay. We know that they were the garrison of Chesterholm for almost two centuries, for they are reported in the

Helmet crest from Vindolanda at the Roman Army Museum in Greenhead. (*Author's collection*)

Notitia Dignitatum: '*Tribunus cohortis quartae Gallorum, at Chesterholm*.' Evidence tells us that Chesterholm fort was abandoned briefly, between AD 245 and 270, and it has been suggested that in that period this unit was moved along the wall to Castlesteads for a short time before being returned to Chesterholm for the rest of their period of service. The datable inscriptions at Chesterholm all fall before or after that period, after all. However, since Cohors II Tungrorum seem to have been in residence at Castlesteads from AD 241 onwards, IV Gallorum cannot reasonably have been there. As such, either the altars were voted by visiting officers from Chesterholm, or perhaps they garrisoned Castlesteads briefly in the Severan era after moving south from Risingham and before taking up residence in Chesterholm.

Thus ends the unit's service record, one of the more complete and straightforward of all the *auxilia* in Britain.

Attested officers and men:

Prefect Quintus Pisentius Justus at Castlehill AD 142–180
Prefect Sulpicius Pudens at Chesterholm
Prefect Pituanius Secundus at Chesterholm
Prefect Quintus Petronius Urbicus at Chesterholm AD 213–235
Prefect Aurelius Mucius at Chesterholm AD 276–282
Prefect Volcacius Hospes at Castlesteads
Prefect Naevius Hilarus
Prefect Titus Flavius Valerianus Papiria Alexander

Prefect Titus Porcius Cornelianus at Marseilles
Prefect Gaius Minicius Italus at Ostia
Prefect Gaius Vindilicius Fontanus in Thracia (Diploma RMD 14)
Emeritus Crotus at Templebrough, second century
Eques Gaius Iulius Valens in Thracia (Diploma RMD 14)
Miles Cintusmus at Templebrough, second century
Melisus at Templebrough, second century
Ve[…] Caecil[…] at Chesterholm
Ca[…]s Ir[…] at Castlesteads

Inscriptions:
Castlehill – altar to the Goddesses of the parade ground (*RIB* 2195)
Castlesteads (Camboglanna) – altar to Jupiter (*RIB* 1979)
Castlesteads (Camboglanna) – altar to Jupiter (*RIB* 1980)
Chesterholm (Vindolanda) – altar to the genius of the praetorium (*RIB* 1685)
Chesterholm (Vindolanda) – altar to Jupiter (*RIB* 1686)
Chesterholm (Vindolanda) – altar to Jupiter (*RIB* 1687)
Chesterholm (Vindolanda) – altar to Jupiter (*RIB* 1688)
Chesterholm (Vindolanda) – dedication to Caracalla (*RIB* 1705)
Chesterholm (Vindolanda) – dedication to Severus Alexander (*RIB* 1706)
Chesterholm (Vindolanda) – dedication to Probus (*RIB* 1710)
Chesterholm (Vindolanda?) – altar to Jupiter (*RIB* 2062)
High Rochester (Bremenium) – building inscription (*RIB* 3491)
Isparta (Saporda?) – dedication of Papiria Alexander (IGR iii 5351)
Marseilles (Massilia) – dedication of Porcius Cornelianus (*ILS* 8852)
Musait-Dunareni (Sacidava) – milestone of Trajan (*AE* 1981, 745)
Ostia – Dedication by Minicius Italus (*AE* 1972, 72)
Risingham (Habitancum) – altar to the divinity of the emperors (*RIB* 1227)
Risingham (Habitancum) – tombstone (*RIB* 1249)
Templebrough – stamped tiles (*RIB* 2458.9–10 and 2472.1–2)
Templebrough – tombstone of Cintusmus (*RIB* 619)
Templebrough – tombstone of Crotus (*RIB* 620)

Cohors V Gallorum (aka Cohors Quintae Gallorum Equitata)

Type: Equitata
Size: Milliary

This milliary (1,000-strong) cohort was raised from the tribes of central Gaul at some point prior to the late Flavian era. They are first attested in Pannonia during the reign of Domitian on two diplomas, and were still in that province in AD 110. Their arrival in Britain pre-dates AD 122, when they are first attested by diploma.

Their early garrison and movements are unknown, for they are first evidenced by epigraphy at Cramond. That fort seems to have had two occupation phases during the Antonine era, but building work suggests that II Tungrorum were the first, while an Antonine milestone lends weight to the second being I Cugernorum. As such, it is possible that V Gallorum formed the garrison during the Severan campaigns of AD 208–211, in which period the altar was raised. Certainly the abbreviation of Jupiter Optimus Maximus to IOM, as seen on this altar, is more common in the later eras.

With the end of the Severan advance and the withdrawals under Caracalla, the unit seems to have been moved to South Shields, which had been expanded to become the main supply base and granary for the northern military. Cohors V Gallorum are attested at South Shields in AD 213 and 222, and the stamped tile *RIB* 2473.9 may confirm a later occupation. If the tile, which is stamped COHVGG, can be translated as Cohors V Gallorum Gordiana, as has been suggested, then that dates the work to AD 238–244. Gordiana is an honorific bestowed upon a number of British units, and so this is quite possible. Though the altar *RIB* 1054 does not name a unit, its dating of AD 211–212 would make it part of this cohort's occupation, and its text ('To the gods the Preservers for the welfare of the Emperor Caesar Marcus Aurelius Antoninus Augustus, Most Great Conqueror of Britain, and of the Emperor Caesar Publius Septimius Geta Augustus, conqueror of Britain, the military unit at Lugudunum paid its vow for their safe return') suggests that the unit raised it in thanks for their return from Severus' Scottish campaigns. This also tells us that Lugdunum was the name of South Shields prior to the fourth century.

There is an interesting possibility opened up in this period. Cohors V Gallorum was a milliary unit, and the barrack blocks in South Shields would be insufficient to house such a unit in full. Moreover, while it is often assumed the unit was purely infantry, the Aquileia inscription names them an *equitata* unit of mixed foot and horse. Once again, South Shields would not be able to reasonably house such a unit. However, there is evidence at Cramond for minor building work that postdates the Severan withdrawal. It is therefore possible that this large unit was divided in around AD 211, with one vexillation moving south to take control of the Severan supply base at South Shields, and another remaining in place at Cramond as a forward garrison, in a similar manner to the outpost forts of Hadrian's Wall. If this was the case, then perhaps the infantry took South Shields, while the cavalry remained in Cramond, where they would be perfect for operations in the easy lands of Lothian.

The Severan granaries of Arbeia. (*Author's collection*)

While the tombstone of Catonius Vindex, found in Rome, explicitly names his unit as Cohors VI Gallorum, no other record of such a unit exists, and it is entirely possible that the extra I is a stonemason's mistake and the deceased was, in fact, a prefect of this unit. As such he has been included on the rolls here.

We can say nothing of the cohort following their time in South Shields, and they were not the final garrison there, for the *Notitia Dignitatum* names that unit as 'Numeri Barcariorum Tigrisiensium', boatmen from the River Tigris, from whom the fort name comes (Arbeia = The place of the Arabs). The changeover between the two units may have taken place sometime around AD 274, when Aurelian had the fort at South Shields reconstructed.

Sword with brass image of Mars, in Arbeia Museum. (*Author's collection*)

Attested officers and men:

Prefect Lucius Minthonius Tertullus at Cramond, late second century
Prefect Gaius Minicius Italus at Aquileia
Prefect Marcus Mactinius Avitus Catonius Vindex, at Rome

Inscriptions:

Aquileia – career inscription (*AE* 1972, 72)
Cramond – altar to Jupiter (*RIB* 2134)

Rome (Roma) – tombstone of Catonius Vindex (*CIL* XI, 1449)
South Shields (Arbeia) – altar to Di Conservatores (*RIB* 1054)
South Shields (Arbeia) – fragment (*RIB* 1059)
South Shields (Arbeia) – dedication to Severus Alexander (*RIB* 1060)
South Shields (Arbeia) – lead seals (*RIB* 2411.100–105)
South Shields (Arbeia) – stamped tiles and bricks (*RIB* 2473.1–9)
South Shields (Arbeia) – dedication to Julia Domna (*RIB* 3272)

Cohors I Hamiorum (aka Cohors Primae Hamiorum Sagittariorum)

Type: Archers
Size: Quingenary

Archers are rare within the framework of the Roman army, and rarer still in Britain. Throughout the period, the most renowned archers generally hail from the eastern Mediterranean, particularly from Crete and Syria. This unit is unique in the known history of Roman Britain, therefore, and Hamian archers take their name from the ancient city of Hama in Syria, which gives us the origin of the unit. Hama was annexed by Rome in 64 BC, and so any units raised from the city postdate this time. We cannot be sure of the date of this unit's founding, and likely they were not raised until the post-Flavian era, once the region had settled following the Jewish Wars. They are mentioned in the Brigetio diploma of 122, which puts them in Britain in that year. Their service prior to this is uncertain, but given the Pannonian diploma, it is possible that after its raising in Syria, the unit was based for a time in Pannonia, and in 122, during Hadrian's reorganization, was moved to Britain.

The cohort was a quingenary unit, consisting of 500 men. As expert archers, it may be that the Hamians were much sought after as local garrisons for their abilities at hunting. The references to hunters along Hadrian's Wall suggest the importance of game in the dietary supply of the local garrisons, and archers would not be numerous among other infantry garrisons.

Our first confirmed garrison dates from circa AD 136, but it is possible to tentatively place this unit, unique in Britain, at the Hadrian's Wall fort of Housesteads (Vercovicium) sometime between 124 and 136, in the reign of Hadrian, when the fort was first constructed. Though we have no certain dates, the tombstone of an archer was found at the fort, very clearly displaying the distinctive look of the Hamians, but with no surviving inscription. This does not confirm the presence of the unit *per se*, especially since the cohort numbered 500 men and Housesteads could comfortably house twice as many, but still it could potentially indicate their presence. Most likely an archer from Carvoran retired to the vicus at Housesteads and was buried there. From the third century until the end of Roman Britain, the fort was occupied by Tungrians, and so if the Hamians were there, it was certainly in the second century, and as we shall see, the unit was rather busy from 136 onwards. Whatever the case, it is highly likely that the Hamian archers were brought to Britain during Hadrian's provincial reorganizations and were immediately based around the northern border and the emperor's new wall.

Around the year 136 they are attested in the Stanegate fort of Carvoran (Magnis), of which they seem to have become the garrison, confirmed by various epigraphy, which tells us they were also there between 197 and 217. There is a question concerning their activity between those times. They may have been the garrison uninterrupted throughout the entire period, but the unit is also attested on two altars at the Antonine Wall fort of Bar Hill. That fort also records a garrison of Baetasii, but since the Antonine Wall seems to have enjoyed two periods of occupation, it is not unreasonable to expect two separate units attested. The question that remains unanswered is whether the entire unit was moved from Carvoran to Bar Hill between 142 and 162, and then returned to their former garrison, or whether they remained the occupying unit of Carvoran, sending a vexillation to either occupy, or bolster the garrison of, Bar Hill. Both Carvoran and Bar Hill are smaller forts, appropriate to a quingenary unit, and so the unit may have simply been divided between the two for a time. One interesting question is raised by the fact that the unit seems to have been commanded by a string of prefects, as is appropriate for a quingenary unit, but with a tribune instead attested once at Carvoran. On this one inscription the unit is not named, but Donatianus is explicitly described as a 'tribune in the post of prefect'. See Cohors I Lingonum also for the rare occurrence of a man of tribune rank acting as a prefect. I have added the man (Marcus Caecilius Donatianus) to the roll of the Hamians based upon the Syriac nature of the inscription he set up, although it is more than possible that his inscription references the Syrian-born empress Julia Domna rather than a deity.

The Hamians seem to have been moved around the northern frontier zone depending upon military policy of the time, likely due to their unique nature, and it is not until after the last great attempt to annex Scotland under Septimius Severus that they seem to move further south. We know that in the post-Severan era the Hamians were gone from Carvoran, for the fort was occupied by a Dalmatian unit in the third century. Their most likely move is suggested by an altar (*RIB* 764) from the fort of Kirkby Thore (Bravoniacum) in the Stainmore Pass. The altar, raised by one Gaius Caledius Florentinus to the Fortune of the Bath House, includes the initials N M S S, which has been tentatively recreated by Mommsen as 'NUMERI MILITVM SYRORVM SAGITTARIORVM' (the unit's name as given on many a diploma). If this is the case, then the Hamians may have decreased in size and been classified as a non-standard *numerus* based at Kirkby Thore, or perhaps this was their base, while other parts of the unit were assigned as missile support elsewhere, which would explain the reduction. The presence of both a dedication at Kirkby Thore raised by a cavalryman of an unknown *ala* (*RIB* 765), and a carved cavalry tombstone, suggest that either I Hamiorum succeeded an *ala* as occupants, or perhaps, if they were a reduced unit, they coexisted with a cavalry force at Kirkby Thore.

The only other late reference (in the *Notitia Dignitatum*) to a unit at that fort is to a 'defending *numerus*', and it may be that this and the Hamians are one and the same unit, particularly since the *Notitia* does not directly mention the Hamians elsewhere. Based on Mommsen's translation and the available evidence, then, it is very possible that after the Severan withdrawal from Scotland, the Hamian archers moved south to Kirkby Thore, where they remained until the close of Roman Britain while supplying archers to other forts as expediency required.

A tombstone from Trier (Augusta Treverorum) which names a member of the unit (Flavius Cornelius Rufinus) who was born in the city of Trallis in Syria but died in Germania, is interesting. It is the unit's only noted link with the region, and so seems likely to be the case of an individual retiree moving to the region after service in Britain. During the reigns of Trajan and Hadrian, a wealth of military diplomas naming British units were issued in Morocco at Sidi Ali bou Djenoun (ancient Banasa) and nearby Volubilis, as well as one just across the water in Iberia at Baelo Claudia. Like northern Britain, that region was very much one of Rome's troubled borders, eternally at risk from nomadic tribes, and so a policy of settling campaign-hardened retirees from Britain to the Moroccan border is far from surprising. It is, however, one of the clearest examples of how Rome served as an exchange of cultures, when a unit from Syria is posted to northern Britain and then largely retired to Morocco. Regardless of this date, the likelihood is still that the actively serving unit remained in Britain, and probably at Kirkby Thore.

Attested officers and men:

Tribune Marcus Caecilius Donatianus at Carvoran, AD 197–217
Prefect Lucius Aelius Caesar Titus Flavius Secundus at Carvoran, AD 136–138
Prefect Licinius Clemens at Carvoran, AD 163–166
Prefect Caristanius Justianus at Bar Hill
Prefect Gaius Iulius Marcellinus at Bar Hill
Prefect [...] Luci, son of Galerius Maximus
Centurion Silvanus at Carvoran
Sabinus at Carvoran

Relief of a Hamian archer from Housesteads, now in the Hancock Museum. (*Author's collection*)

Replica Hamian Helmet on display at the Roman Army Museum in Greenhead. (*Author's collection*)

Marcus Mussius Concessus, AD 132
Gaius Caledius Florentinus at Kirkby Thore
Flavius Cornelius Rufinus

Inscriptions:
Housesteads (Vercovicium) – tombstone of an unknown archer
Carvoran (Magnis) – altar to the emperor's fortune (*RIB* 1778)
Carvoran (Magnis) – altar to Hammia (*RIB* 1780)
Carvoran (Magnis) – altar to the Syrian goddess (*RIB* 1792)
Carvoran (Magnis) – altar (*RIB* 1809)
Carvoran (Magnis) – poetic inscription to Virgo Caelestis (*RIB* 1791)
Carvoran (Magnis) – unidentified inscription (*RIB* 1810)
Carvoran (Magnis) – centurial stone of Silvanus (*RIB* 1820)
Bar Hill – altar to Mars Camulus (*RIB* 2166)
Bar Hill – altar to Silvanus (*RIB* 2167)
Bar Hill – tombstone (*RIB* 2172)
Kirkby Thore (Bravoniacum) – altar to Fortuna Balnearis (*RIB* 764)
Lliria, Spain – tombstone (*CIL* II, 3791)

Cohors I Hispanorum (aka Cohors Primae Aelia Hispanorum Milliaria Equitata)

Type: Equitata
Size: Various

This cohort raises one major difficulty. On the bulk of the evidence, it is easy to assign all finds to one unit, but the last relevant diploma of AD 178 includes the listing '*et I Aelia Hispanorvm et I fida Vardvllorvm et I Celtiberorvm et III Lingonvm et I Hispanorvm et…*' This indicates rather clearly that there were two units with the same name and number both operating in the province, the only real distinction being that the honorific 'Aelia' had been applied to one of them. Our early evidence could fit either unit, as before the arrival of an Aelian emperor there is no clear distinction possible, although as noted below, there is reason to assign the early to the non-Aelian cohort.

The Ager Nerviorum diploma of AD 98 names a Cohors I Hispanorum Asturum, and the *Notitia Dignitatum* also names a I Asturum based at Aesica, though we should beware of connecting the two. Since there is ample epigraphic evidence at Aesica for Cohors II Asturum, it is almost certain that the name of the unit on the *Notitia* is a transcription error and should refer to II Asturum. That being the case, then the Ager Nerviorum diploma would refer to one of these units and, when combined with the Ardoch epigraphy (see below), confirms a presence in Britain by AD 98. The units are then attested in the province on diplomas until AD 178.

Of the units' origin and early service we can say little, other than that based upon diploma evidence they were raised from among the Astures in Spain, and that at least one of them was based at some unknown location in Britain by AD 98. Our first evidence in the province comes from a tombstone found at Ardoch in Scotland, which references I Hispanorum and has been dated to the Flavian era by the use of '*Dis Manibus*', rather than the later common abbreviation 'D.M.' This makes it very likely that one unit was brought to Britain by the Flavians and served in the campaigns of Agricola in Scotland, culminating in the Battle of Mons Graupius in AD 83. The units' movements during the late Flavian, Nervian and Trajanic eras are uncertain.

I Aelia Hispanorum

At this point we can perhaps start to divide out the units, and it may be that it is under Hadrian that this unit is first attested on the island, with all prior evidence pointing to the other cohort. Our first epigraphic evidence comes from Maryport and is of Hadrianic date. It would appear that although initially a quingenary cohort, the unit had become milliary by AD 122, and Maryport was certainly of a size to hold the larger unit. Their time at the fort, which was curtailed before the Severan era, has produced one of the best epigraphic records of any fort or unit in Britain. We are aware of a number of commanders thanks to the finds of a large number of altars, and one officer in particular has drawn much discussion. Marcus Maenius Agrippa is noted on *ILS* 2735 as having been sent by Hadrian on the 'British Expedition' (*electo a divo Hadriano et misso in expeditionem Britannicam*). This may refer to the rebellion of AD 119–120, and so Agrippa was possibly assigned then. In the same inscription, he is noted as having hosted the emperor. Agrippa came from the town of Camerinum in Italy, and Hadrian visited that area in AD 127, so this seems likely to be the occasion mentioned.

The unit acquired its honorific 'Aelia' either under Hadrian or his successor, both of whom bore that name. If it was under Hadrian, then perhaps it was with Agrippa during that rebellion that the honorific was acquired. However, based on the fact that none of the Hadrianic altars of Maryport mention Aelia, an Antonine event is perhaps preferable. It is likely that the 'Aelia' honorific was acquired during the campaigns in Scotland under Antoninus Pius (AD 140–142). Cohors I Delmatarum took control of Maryport in the Antonine era, and I Aelia Hispanorum moved on.

Our next information about the cohort shows them taking control of the fort of Netherby (Castra Exploratorum), one of the outpost forts of Hadrian's Wall, which had previously been occupied by Cohors I Nervana. It is likely that I Aelia Hispanorum took control of Netherby during Septimius Severus' reorganization in preparation for his invasion of Scotland in AD 208. Certainly all their epigraphy at the fort is Severan, and often involves construction. The fort was expanded at this time to a larger size, so perhaps their arrival coincided with the fort's expansion to house them. Interestingly, building work here seems to have involved the legions too, for detachments of both II Augusta and XX Valeria Victrix are commemorated alongside I Hispana on at least one dedication from the reign of Elagabalus. The unit's links with the Severans are given added weight by the addition of the honorific 'Severiana Alexandriana' in AD 222.

We cannot be certain of the fate of I Aelia Hispanorum after Netherby, which may have been reduced to a fortlet and occupied by scouts. The unit that appears on the *Notitia Dignitatum* omits the Aelia, which is unusual, and so almost certainly points to the other unit. As such, our last evidence of I Aelia Hispanorum in Britain is early third-century Netherby.

I Hispanorum

Following AD 222, our only knowledge comes from the early fifth-century *Notitia Dignitatum*, which places the unit at one 'Axeloduno'. This fort is neither Netherby, nor Carlisle (Uxellodunum), since that also appears on the same list. We have no idea of the modern location of this Axeloduno that saw the last phase of the unit's service. Following the sequence in the *Notitia*, Axeloduno falls between Concavata and Gabrosentum, the forts of Drumburgh and Moresby respectively, possibly indicating Kirkbride, Blennerhassett or Beckfoot forts, for none of which we know the Latin name. Whatever the case, I Hispanorum seem to have been serving in Britain to the very end. As such, the evidence seems to suggest:

- I Hispanorum serving in the Flavian campaigns, in Scotland, then vanishing from the record until they are recorded at the end of Roman Britain in the unknown fort of Axeloduno.
- I Aelia Hispanorum first entering the records under Hadrian, manning Maryport before moving to Netherby, following which they disappear from the record.
- The only record of both units serving simultaneously being the diploma of AD 178. Any discovery that there was an error in the 178 diploma, of course, would allow these two units to be easily conflated.

The impressive defences of Ardoch fort. (*Author's collection*)

Notes:

1. The almost illegible altar *RIB* 836 has been assigned to I Aelia Hispanorum on stylistic grounds.
2. Cornelius Peregrinus is not assigned to a specific unit on the inscription, but the altar has been dated to the second century, when I Hispanorum was very likely the garrison.
3. M. Censorius Cornelianus is a temporary commander, from the inscription text. He is a centurion of Legio X Fretensis, and as praepositus, is the *acting* commander of the unit. The reason for this is sadly unknown, but might be interesting, given that his legion was based in Jerusalem, with the only known exception in our period being its service in Marcus Aurelius's Marcomannic Wars.

Attested officers and men:

Tribune Cornelius Peregrinus at Maryport
Tribune Gaius Caballius Priscus at Maryport
Tribune Marcus Maenius Agrippa at Maryport
Tribune Marcus Aurelius Salvius at Netherby, AD 219
Tribune [...] Maximus at Netherby, AD 218
Prefect Lucius Antistius Lupus Verianus at Maryport
Prefect Publius Cornelius Ur[...] at Maryport
Prefect Helstrius Novellus at Maryport
Prefect Lucius Cammius Maximus at Maryport, AD 133–135
Praepositus (Centurion) Marcus Censorius Cornelianus at Maryport
Centurion Ammonius, late first century

Inscriptions:

Ardoch (Alauna) – funeral inscription of Ammonius (*RIB* 2213)
Camerino (Camerinum) – inscription of Marcus Maenius Agrippa (*ILS* 2735)
Maryport (Alauna Carvetiorum) – altar to genius loci, Fortuna, Rome and good fate (*RIB* 812)
Maryport (Alauna Carvetiorum) – altar to Jupiter (*RIB* 814)
Maryport (Alauna Carvetiorum) – altar to Jupiter (*RIB* 815)
Maryport (Alauna Carvetiorum) – altar to Jupiter (*RIB* 816)

The 'Netherby Genius', now in Tullie House Museum. (*Author's collection*)

Maryport (Alauna Carvetiorum) – altar to Jupiter (*RIB* 817)
Maryport (Alauna Carvetiorum) – altar to Jupiter (*RIB* 818)
Maryport (Alauna Carvetiorum) – altar to Jupiter (*RIB* 819)
Maryport (Alauna Carvetiorum) – altar to Jupiter (*RIB* 820)
Maryport (Alauna Carvetiorum) – altar to Jupiter (*RIB* 821)
Maryport (Alauna Carvetiorum) – altar to Jupiter (*RIB* 822)
Maryport (Alauna Carvetiorum) – altar to Jupiter (*RIB* 823)
Maryport (Alauna Carvetiorum) – altar to Jupiter (*RIB* 824)
Maryport (Alauna Carvetiorum) – altar to Jupiter (*RIB* 825)
Maryport (Alauna Carvetiorum) – altar to Jupiter (*RIB* 826)
Maryport (Alauna Carvetiorum) – altar to Jupiter (*RIB* 827)
Maryport (Alauna Carvetiorum) – altar to Jupiter (*RIB* 828)
Maryport (Alauna Carvetiorum) – altar to Jupiter (*RIB* 829)
Maryport (Alauna Carvetiorum) – altar (*RIB* 836)
Maryport (Alauna Carvetiorum) – altar to Volcanus (*RIB* 846)
Maryport (Alauna Carvetiorum) – building dedication (*RIB* 855)
Netherby (Castra Exploratorum) – altar to Fortuna Conservatrix (*RIB* 968)
Netherby (Castra Exploratorum) – dedication to Julia Domna (*RIB* 976)
Netherby (Castra Exploratorum) – building dedication (*RIB* 977)
Netherby (Castra Exploratorum) – dedication to Severus Alexander (*RIB* 978)
Netherby (Castra Exploratorum) – dedication to Severus Alexander (*RIB* 979)
Netherby (Castra Exploratorum) – building dedication (*RIB* 980)
Thyna (Thenae) – career inscription (*AE* 1949, 37)

Cohors I Lingonum (aka Cohors Primae Lingonum Equitata)

Type: Equitata
Size: Quingenary

Five cohorts are known to have been drawn from the Lingones in eastern France in the same period, possibly following the Civilis Rrevolt of AD 69/70. Of those, Cohors V is only referenced serving in Dacia, while the other four are all attested in Britain. Cohors I Lingonum first appears in the province on the diploma of AD 105, but was likely on the island from the Flavian period. Our first epigraphic evidence is from the fort of High Rochester and, although that evidence is dated to the Antonine period, since the fort existed from the Agricolan campaigns onward it is possible that I Lingonum was the garrison throughout that period. The fort was rebuilt in stone around AD 139, and the evidence suggests that I Lingonum were involved in the construction.

Our next evidence for the unit comes from the fort of Lanchester and can be dated to AD 238–244. Though we have no concrete evidence for the date they left High Rochester, it is interesting to note that the previous occupants of Lanchester (I Vardullorum) moved to High Rochester prior to AD 213, and so it must have been before then. Since I Vardullorum and I Lingonum swap garrisons, it seems highly likely that both units were drawn from their forts in AD 208, forming part of the army Septimius Severus took on his Scottish campaign. When that war ended in AD 211 and Caracalla withdrew the forces, the two units were sent to occupy each other's former home.

By AD 238, then, I Lingonum were in Lanchester, attested by building inscriptions, and seem to have acquired the honorific 'Gordiana'. During their stay there, they were commanded by a string of prefects, and, oddly, once by a tribune. This is not an unknown situation, and is also recorded at Carvoran where a man of a tribune's rank served as a prefect, and this seems to be an innovation of the third century onwards. The Stanhope altar was likely raised by a prefect commanding at nearby Lanchester following a successful hunting trip.

We cannot be certain of the fate of I Lingonum, though it is possible they remained the garrison of Lanchester until the end. The *Notitia Dignitatum* tells us of '*Praefectus numeri Longovicanorum, at Longovicio*', rather vaguely. The identity of this unit, the '*Numerus of Longovicium*', cannot be divined from the name, and so it is possible that this is the same unit as I Lingonum, naturalized to their location over two centuries.

Attested officers and men:
Tribune Claudius Epaphroditus Claudianus
Prefect Fulvius Felix at Lanchester, early third century
Prefect Marcus Aurelius Quirinus at Stanhope, AD 238–244
Prefect Gaius Caesidius Dexter at Macerata Feltria

Dedication from High Rochester, now in the Hancock Museum. (*Author's collection*)

Inscriptions:
High Rochester (Bremenium) – Antonine dedication (*RIB* 1276)
Lanchester (Longovicium) – dedication to the *genius* of the *praetorium* (*RIB* 1075)
Lanchester (Longovicium) – dedication to Gordian III (*RIB* 1091)
Lanchester (Longovicium) – dedication to Gordian III (*RIB* 1092)
Lanchester (Longovicium) – altar to Jupiter (*RIB* 3262)
Macerata Feltria (Pitinum Pisaurense) – dedication (*CIL* XI, 6033)
Stanhope – altar to Silvanus (*RIB* 1042)

Cohors II Lingonum (aka Cohors II Lingonum Equitata)

Type: Equitata
Size: Quingenary

This unit was raised among the Lingones of northeastern France at some point prior to the reign of Trajan. Given the name of the recipient of the AD 127 diploma – one Itaxa, son of Stamilla the Dacian – we can tentatively build an early record. If Itaxa had served twenty-five years in AD 127, his father may have also served in the *auxilia*, which could easily put his recruitment around AD 88, suggesting he might have joined the Roman army following Domitian's Dacian War. If this is the case, then that suggests that II Lingonum was serving somewhere in that region in the Flavian era, possibly even involved in the war of AD 86–88. Certainly, since the unit is first attested in Britain in a diploma of AD 98, Itaxa and his father cannot have joined up as a result of Trajan's Dacian Wars.

The unit's first activity in Britain under Trajan is unknown, and their epigraphy at the coastal station of Moresby is undated. However, given that they were elsewhere in the later second century, and were not the garrison by the time of the *Notita Dignitatum*, it is possible that Cohors II Lingonum were the first garrison at Moresby, created as part of Hadrian's defensive system. Moresby was then occupied by Cohors II Thracum, probably on the latter's withdrawal from the Antonine Wall. Thus, II Lingonum were likely replaced in that fort in AD 160 or thereabouts, and moved to a new garrison.

The dates then fit smoothly for a new posting at the fort of Ilkley. An altar at Ilkley was raised to the emperors Marcus Aurelius and Lucius Verus, which puts their presence there between AD 161 and 169. Lead seals found at Brough-under-Stainmore cannot be taken as proof of residence, and so the only remaining evidence we have to consider is the fifth-century *Notitia Dignitatum*. That work records '*Tribunus cohortis secundae Lingonum, at Congavata*', the fort of Drumburgh on Hadrian's Wall. When their move from Ilkley to Drumburgh occurred we cannot even conjecture, other than that it occurred between the reign of Marcus Aurelius and the end of Roman government in Britain.

Roman stonework from the fort built into Drumburgh castle. (*Author's collection*)

Attested officers and men:
Prefect Cloduis Fronto at Ilkley
Prefect Caecilius Lucanus at Ilkley, AD 161–169
Prefect Valerius Lupercus at Moresby
Prefect Gaius Hedius Verus in Britannia (*CIL* XI, 612 and diploma *ZPE* 117, 274)
Commander Gaius Pompeius Saturninus at Moresby
Miles Itaxa

Inscriptions:
Fossombrone (Forum Sempronii) – career inscription of Hedius Verus (*CIL* XI, 6123)
Ilkley (Verbeia?) – altar to Verbeia (*RIB* 635)
Ilkley (Verbeia?) – inscription to the emperors (*RIB* 636)
Moresby (Gabrosentum) – altar to Silvanus (*RIB* 798)
Moresby (Gabrosentum) – inscribed panel (*RIB* 800)

Cohors IV Lingonum (aka Cohors IV Lingonum Equitata)

Type: Equitata
Size: Quingenary

Raised among the Lingones in what is now northern Burgundy, the date of this unit's founding is unknown. They first enter the historical record on a diploma of 105, placing them in Britain in that year, and diplomas continue to show them in the province until at least AD 158. Indeed, we have only British activity throughout the unit's history, barring two retired prefects living elsewhere.

Where the unit was originally based in Britain we cannot say, though at the end of the second century they seem to have replaced Cohors II Nerviorum as the garrison of Wallsend fort, whence all British epigraphy comes, including numerous stamped tiles. It seems likely that they took control of the fort at the start of the Severan campaigns in the north. They are attested on a stone that seemingly refers to building work on the vallum, way to the west at Irthington. This could be taken as their part in the initial construction of the wall, or perhaps in its reorganization under Severus. An altar at Wallsend raised by the unit shows that it was commanded at that time by a centurion of Legio II Augusta. Again, this could reference the wall's original construction, but since we have confirmed third and fourth-century occupation for the fort, and we know II Nerviorum were there previously, it seems to fit well with the Severan work on the wall. Severus's campaigns are one of the few occasions when Legio II is known to have been active in the north.

Two of the Wallsend altars have been dated to the third century, placing Cohors IV Lingonum there then, and we know they remained there until the end of Roman Britain, for the *Notitia Dignitatum* tells us of '*Tribunus cohortis quartae Lingonum, at Segeduno.*'

The career inscription of Statius Priscus in Rome clearly references this unit. The Tarragona one was initially read as III Lingonum, but has been more recently reassessed and assigned to IIII Lingonum, and so Lucretius has been included in the unit roster.

Attested officers and men:
Prefect Aelius Rufus at Wallsend, third century
Prefect Cornelius Celer at Wallsend, third century
Prefect Didius Severus at Wallsend
Prefect Marcus Statius Priscus at Rome
Prefect Marcus Lucretius Quir. Peregrinus at Tarragona
Centurion Iulius Honoratus at Wallsend

Reconstructed bathhouse at Wallsend. (*Author's collection*)

Inscriptions:
Irthington – building inscription (*RIB* 2014)
Rome (Roma) – career inscription of M. Statius Priscus (*ILS* 1092)
Tarragona (Tarraco) – inscription of Lucretius Peregrinus (*AE* 1961, 330)
Wallsend (Segedunum) – altar to Jupiter (*RIB* 1299)
Wallsend (Segedunum) – altar to Jupiter (*RIB* 1300)
Wallsend (Segedunum) – altar to Jupiter (*RIB* 1301)
Wallsend (Segudenum) – fragmentary dedication (*RIB* 1302)
Wallsend (Segudenum) – stamped tiles (*RIB* 2476)

Cohors I Nerviorum (aka Cohors Primae Augusta Nerviorum/ Nervana Germanorum)

Type: Equitata
Size: Milliary

British diplomas and epigraphy give us a range of very similar unit names for this cohort. There has been considerable discussion as to whether they all refer to the same unit, or to two, three, or even five different units. The given names present the following variations: Nerviorum, Nerviana, Nervia, Nervana. The name refers to the tribe of the Nervii, who occupied what is now northeastern

Belgium, and not, as has been conjectured, the emperor Nerva. I have chosen to present all these possibilities as one unit on the basis that all variations of the name including an 'i' appear on diplomas, while all those without an 'i' belong to inscriptions, suggesting that they refer to the same unit. And given that diplomas refer to 'Nervia German', and inscriptions to 'Nervana Germanor', we can assume a singular unit raised from the Nervii and later acquiring the 'German' epithet.

The unit was likely raised as a quingenary one from among the Nervii, at some point in the mid-first century. No record of the unit appears before their service in Britain, recorded initially on the diploma of AD 105. Our first inscription is undated but likely comes from the late first century, from the fort of Caer Gai in Wales. Davies dates construction of Caer Gai to Frontinus circa AD 77, and it is possible that the newly raised Cohors I Nerviorum was its first garrison, especially given that the inscription discovered there states only 'COH I NER', without the honorific Augusta or the epithet Germanorum, which were to come later.

There is an odd question, then, raised by a diploma of AD 107 from Mauretania Caesariensis, which states that a 'I AUGUSTA NERVIANA VELOX' was serving in that province. Given that two years earlier they were in Wales, and by AD 122 they were in Britain again, there are two possible explanations for this. It is possible that between AD 105 and 107 the unit was transferred to Caesariensis to recruit, and that it was at this time it became a milliary cohort, bolstered by 500 Moors. Another possible explanation, of course, is that I Augusta Nerviana Velox is an entirely different unit. Whatever the case, whether the unit was moved briefly to another province and back or not, the next evidence comes from Birrens, and shows changes in the unit.

The unit's stay at Birrens has been dated to AD 142–156, based on construction dates and finds. Birrens had already existed as a Hadrianic fort, but was only large enough for a quingenary unit. Around AD 142, it was rebuilt as a much larger fort. This suggests that I Nerviorum had become a milliary unit by now. Perhaps this was the result of a brief recruiting drive in Mauretania, or perhaps, since from the diploma of AD 122 onwards they have acquired the epithet Germanorum, they had been bolstered by an influx of Germanic recruits. Their stay at Birrens seems to have been entirely Antonine, during the reoccupation of Scotland, when the advance forts of Hadrian's Wall would have become important. They were replaced at Birrens by Cohors II Tungrorum only five years before the fort was abandoned.

An altar that attests 'Coh I Nervana' was recorded at Netherby (*RIB* 966) and has been linked to three other altars at the fort of Bewcastle, but since none of the Bewcastle altars name a unit, any association of I Nerviorum with that

The ramparts of Caer Gai. (*Author's collection*)

fort is speculation at best. However, a late second-century date applied to the Netherby altar also makes it unlikely to be connected with their stay 10 miles away at Birrens. The answer likely comes from their next move.

The cohort is attested on two inscriptions from Burgh-by-Sands, as a milliary unit, and it seems likely that this garrison followed their move from Birrens in AD 156. Interestingly, one of the inscriptions, among the latest for the unit, is a dedication to Hercules, the same deity named on our first inscription from Caer Gai, suggesting a possible connection between the unit and the god throughout their service. Notably, Burgh-by-Sands is less than 10 miles from Netherby, and could therefore be the reason for that lone altar's presence there, courtesy of a visiting officer.

After the two Burgh-by-Sands inscriptions, we have no further record of the unit, but the *Notitia Dignitatum* records the last garrison of that fort as the 'Numerus Maurorum Aurelianorum', an irregular unit from Mauretania. Given the possibility of I Nerviorum having been briefly posted for recruitment in Mauretania, it is faintly possible that all these units are the same one, and that by the end of the fourth century, the 'Germanorum' had been dropped in favour of a Moorish connection. Without further evidence coming to light, all we can do is speculate.

Attested officers and men:
Tribune Paternius Maternus at Netherby(?), late second century
Tribune Lucius Faenius Felix at Birrens, AD 142–156
Tribune Publius Tuscilius [...]asinianus at Burgh-by-Sands
Miles Iulius at Caer Gai

Inscriptions:
Caer Gai – dedication to Hercules (*RIB* 418)
Birrens (Blatobulgium) – altar to Fortuna (*RIB* 2093)
Birrens (Blatobulgium) – altar to Jupiter (*RIB* 2097)
Burgh-by-Sands (Aballava) – Altar to Hercules (*RIB* 2040)
Burgh-by-Sands (Aballava) – altar to Jupiter (*RIB* 2041)
Netherby (Castra Explioratorum) – altar to Cocidius (*RIB* 966)

Cohors II Nerviorum (aka Cohors II Nerviorum Civium Romanorum)

Type: Infantry
Size: Quingenary

This infantry unit was raised among the fierce Nervii with whom Caesar was so impressed, probably in the aftermath of the Civilis Revolt of AD 69. They were likely brought to Britain by Petillius Cerialis in AD 71 and served throughout the campaigns in the north and in Scotland. Our first confirmed evidence of them in Britain comes from the diploma of AD 98, with sporadic diplomas continuing until AD 145.

We cannot be certain of the unit's location throughout the Trajanic and Hadrianic eras. The apparent first evidence we have for them comes from the fort of Wallsend, and they have been presumed to be the initial garrison there, based on the second-century *RIB* 1303. However, the first fort phase of Wallsend included four cavalry barracks, and all the Nervian units in Britain were infantry, so it seems unlikely that this was the case. Instead, perhaps we are better presuming a later second-century date for their presence here, perhaps Antonine. Altar *RIB* 1302 from Wallsend does not name a unit, but if it can be taken to be II Nerviorum, then we have another named prefect in Didius Severus.

Their presence is suggested at Chesterholm by a third-century altar. Although they are attested elsewhere by AD 213, it is possible that the unit was moved to Chesterholm in the post-Antonine period, perhaps as part of the preparations for Severus' campaigns. It is tempting to see them moved there in AD 208 as the northern military was rearranged for the Scottish war, and the undated building works at both Risingham and High Rochester by vexillations of II Nerviorum

The multiple defensive ditch system of Whitley Castle. (*Author's collection*)

carried out from a base at Chesterholm, just 15 and 20 miles away respectively. Similarly, the altar found at Carrawburgh was raised by a vexillation of the unit, and therefore cannot be taken as evidence of their presence as a garrison. They likely donated the altar while in garrison at either Wallsend or Chesterholm.

Whatever the case for their dates at Chesterholm, we can be certain they were at Whitley Castle by AD 213, when they raised an inscription to Caracalla. On this inscription they are noted for the first time with the addition to their name of 'Civium Romanorum', and so it seems likely they gained the grant of citizenship in the Severan era prior to that date. Their movements from AD 213 are unknown, but because they are the only known garrison of Whitley Castle, and that fort is not referenced in the *Notitia Dignitatum*, it is quite possible that they remained the garrison there until the end of Roman Britain.

Attested officers and men:
Prefect Decimus Caerellius Victor at Chesterholm, third century
Gaius[...] at Whitley Castle

Inscriptions:
Carrawburgh (Brocolitita) – altar to *genius loci* (*RIB* 1538)
Chesterholm (Vindolanda) – altar to Cocidius (*RIB* 1683)

High Rochester (Bremenium) – building inscription (*RIB* 3491)
Risingham (Habitancum) – building inscription (*RIB* 1240)
Wallsend (Segedunum) – dedication to Mercury (*RIB* 1303)
Whitley Castle (Epiacum) – altar to Apollo (*RIB* 1198)
Whitley Castle (Epiacum) – dedication to Caracalla (*RIB* 1202)

Cohors VI Nerviorum (aka Cohors Quartae Nerviorum)

Type: Infantry
Size: Quingenary

Like the other Nervian units serving in Britain, Cohors VI Nerviorum was likely raised in the aftermath of the Civilis Revolt and brought to the island in AD 71 by Petillius Cerialis, and therefore probably took part in the Flavian campaigns in northern Britain.

Our first certain evidence of their presence comes from the diploma of AD 122, and we can hazard an educated guess at their movements at this time. They are mentioned on a building inscription from Carvoran, as well as upon an altar from nearby Great Chesters. It is possible, then, that the unit was the first occupant of either of these two forts in the Hadrianic era, involved in building the wall. Evidence favours Great Chesters as their home, the unit dedicating an altar in a fort they occupied while constructing the next fort only 3 miles away.

Following the accession of Antoninus Pius and the advance into Scotland, VI Nerviorum are attested at the Antonine Wall fort of Rough Castle on both an altar and a building inscription. Interestingly, they are given at that time as under the command of a centurion from Legio XX Valeria Victrix. The reason for this is unknown, but it is possible that either there was a lacuna in command for a time, or that the cohort was split between two sites during construction, and the one at Rough Castle was given to the command of a legionary centurion. Whatever the case, at least a vexillation of VI Nerviorum, if not the whole unit, was based at Rough Castle sometime between AD 142 and 165.

By the Severan era, the unit had been moved to Brough-by-Bainbridge. Epigraphy puts them there in AD 205, during the second consulship of Caracalla and Geta, and they are well attested for a time. A lead seal of Cohors II Asturum found at Brough-by-Bainbridge should not be taken as evidence of occupation, and likely belonged to a message sent from another fort to this one. Indeed, though all our epigraphy from the fort dates to the Severan era, the *Notitia Dignitatum* tells us of '*Tribunus cohortis sextae Nerviorum, at Virosido*', which suggested they remained the garrison of Brough-by-Bainbridge from AD 205 at the latest until the end of Roman Britain.

It is worth noting that among the other Nervian cohorts we have just two building inscriptions, both mentioning only a vexillation, while Cohors VI appear to have been prolific builders.

Attested officers and men:
Prefect Gaius Iulius Barbarus at Great Chesters
Prefect Lucius Vinicius Pius at Brough-by-Bainbridge, AD 205–208
Centurion Iulius Martinus at Brough-by-Bainbridge
Centurion Flavius Betto (Legio XX)

Building inscription from Brough-by-Bainbridge. (*Author's collection*)

Inscriptions:
Brough-by-Bainbridge (Virosidium) – Severan dedication (*RIB* 722)
Brough-by-Bainbridge (Virosidium) – dedication (*RIB* 723)
Brough-by-Bainbridge (Virosidium) – building inscription (*RIB* 3215)
Brough-by-Bainbridge (Virosidium) – building inscription (*RIB* 3216)
Carvoran (Magnis) – building inscription (*AE* 1988, 836)
Great Chesters (Aesica) – altar to Imperial Victory (*RIB* 1731)
Rough Castle – altar to Victory (*RIB* 2144)
Rough Castle – building inscription (*RIB* 2145)

Cohors II Pannoniorum (aka Cohors Secundae Pannoniorum)

Type: Infantry
Size: Quingenary

There is a paucity of information concerning this unit. If we were to take only the evidence that distinctly says 'II Pannoniorum', then we have only one piece of epigraphy: the building inscription from Beckfoot. We know Cohors II Pannoniorum were in Britain at least between AD 105 and 124 from diplomas, and we can realistically put them as the first garrison of Beckfoot in the Hadrianic era, constructing and occupying the fort as part of the new integrated wall system. They are named on two lead seals from Chesterholm, but these were likely attached to documents sent from Beckfoot. The seals and the inscription from Beckfoot are our only British evidence.

An inscription from Benevento names a prefect of the unit, though it is not certain the unit was in Britain during his tenure. He has been added to the list regardless, as has the tombstone of Dagvalda from Cawfields milecastle, for the only Pannonian unit in Britain that can be confirmed is this one, and so it seems likely that Dagvalda was a soldier of II Pannoniorum. Inscriptions found at Apulum (*CIL* III, 12631) and Alba Fucens (*CIL* IX, 3924) both seem to denote a Cohors IIII Pannoniorum, and so have been omitted here.

Dagvalda tombstone, now in Chesters Museum. (*Photo courtesy of Pete Savin, with many thanks*)

Inscription from Beckfoot, now in the Senhouse Museum. (*Author's collection*)

It has been conjectured that the unit was based in Dacia following their service in Britain, with a change in the name, for a stamped pot from Buciumi in Romania bears the letters 'COH II N PA'. This has been read as Cohors II Nova Pannoniorum, although it more likely refers to II Nervia Pacensis, which is attested in stamped tiles at Buciumi and is noted in Dacia on diplomas. As such, if we disregard this, we are once more left with no record of II Pannoniorum prior to their presence in Britain in AD 105, and after their last appearance on a diploma there in AD 124. Much of the unit's service remains a mystery.

Attested officers and men:
Prefect Gaius Oclatius Modestus at Benevento
Miles Dagvalda at Cawfields milecastle

Inscriptions:
Beckfoot (Bibra) – building inscription (*RIB* 880)
Benevento (Beneventum) – career inscription (*CIL* IX, 1619)
Cawfields milecastle – tombstone of Dagvalda (*RIB* 1667)

Cohors I Sunucorum (aka Cohors I Sunicorum, Cohors Primae Sunucorum)

Type: Infantry
Size: Quingenary

For this unit, drawn from a German tribe, we have that rarest of things, written evidence. Tacitus, in his Histories (book IV, 66) tells us:

> Civilis [...] decided to try to win over the neighbouring peoples, or, if they refused, to attack them. He had already gained the Sunuci and had organized their young men into companies of infantry.

Thus it seems extremely likely that this unit is one of those mentioned by Tacitus as revolting against Rome with Civilis. Following the suppression of the revolt, the company was probably officially added to the muster roll of the Roman army and brought to Britain as an auxiliary unit by Petillius Cerialis. It is interesting to note that what may be their first evidence in the province, a brick stamped with 'LEG XX VV' found at Holt, is also inscribed with '*Iulius Aventinus miles cortis prima Sunicorum.*' It is tempting to think that, having been involved in a revolt against Rome, the unit was initially placed under the supervision of the local legion. There is, however, a more realistic probability.

Cohors I Sunucorum is attested in the fort of Caernarvon, and, indeed, is the only unit named there at all. It may be that I Sunucorum arrived in Britain and were sent straight to Wales, where they settled into the new Agricolan fort of Caernarvon and remained there for the rest of their time on the island. If so, they would be one of the most settled units in the island's Roman history. Holt is less than 70 miles from Caernarvon via the main Agricolan route across the north coast of Wales, and was a massive tile-works of Legio XX, and it is therefore quite possible that the brick inscription from there was left by a soldier from Caernarvon, visiting to secure tiles for his commander. A paving slab incised with COR found in the bathhouse at Prestatyn, between Caernarvon and Holt, probably refers to this unit.

As well as that of this busy soldier, we have two other names to associate with the unit, both drawn from the Stannington diploma of AD 124, which was given to a retiring soldier of I Sunucorum. Though the name of the retiree is sadly missing, it names his father, one [...] Albani of the Sunuci, who probably also served, as well as his commander, M. Iulius Claudianus.

In addition to that diploma, three others confirm the presence of the unit in Britain between AD 122 and 127, and the inscription from Caernarvon places

The strongroom of the fort of Caernarvon. (*Author's collection*)

them there in the Severan era (AD 198–209). Though the *Notitia Dignitatum* mentions neither this cohort nor the fort of Caernarvon, the location given in the seventh-century Ravenna Cosmography, along with archaeological evidence, tells us that the fort was still occupied at the end of Roman government on the island, so it is therefore quite feasible that I Sunucorum remained its garrison until the very end. An alternative possibility is that the unit was transferred abroad, taking the name of the Caernarvon fortress (Segontium) with them, for an *auxilia palatina* unit named Seguntienses is listed among the troops of the Comes Illyricum on the *Notitia Dignitatum*.

Attested officers and men:
Commander Marcus Iunius Claudianus
Miles Iulius Aventinus, at Holt
[...] Albani

Inscriptions:
Caernarvon (Segontium) – commemorative slab (*RIB* 430)
Holt (Bovium) – stamped brick (*RIB* 2491.96)
Prestatyn – fragment (*RIB* 3148)

Cohors I Thracum (aka Cohors Primae Thracum Equitata Civium Romanorum)

Type: Equitata
Size: Quingenary

Cohors I Thracum could be one of the oldest units to have set foot in Britain. They may well have been raised by Tiberius, and there is evidence of their presence in Germany early on, based at Remagen. They are attested in that province in diplomas until AD 82, based upon the assumption there were not two units with the same name serving close to one another.

If the Wroxeter tombstone missing a number does refer to I Thracum, and not a different unit number, that is still no indication of an early presence on the island. Despite bearing a Claudian or Neronian name, the deceased could have come to Britain at the end of his service and died in Wroxeter. There is no specific recruitment age for the *auxilia*, and therefore at 57, Tiberius Claudius Tirintius may have taken his name in the reign of Claudius or Nero, and still been serving his twenty-five years when the unit was transferred to Britain, post AD 82. It is worth noting here that this is only the most *likely* sequence of events, for the confusion surrounding the possibility of several units named I Thracum makes it difficult to say anything for certain.

We cannot suggest that the unit had any association with Wroxeter other than one of its veterans being buried there. It may point to a late first-century garrison somewhere in the west, perhaps in Wales, however. One thing we can be certain of is that the unit was in Britain by AD 122, from diploma evidence. Their early locations and work in the province cannot be confirmed, though a building inscription found in the vallum of Hadrian's Wall between Newcastle and Benwell suggests that I Thracum was involved in the wall's construction. They may or may not have had a Hadrianic garrison site along the wall, but other than the tombstone at Wroxeter and the slab from the wall, the first datable evidence we have of the cohort comes from Bowes, and is distinctly Severan in date. It would appear that the unit was based at Bowes by AD 197/198. Interestingly, the Spanish cavalry garrison of Binchester, just 20 miles away, are recorded as having helped the Thracians in construction work at Bowes, rebuilding a bath house 'burned by violent fire' (*RIB* 730) and apparently with other work too (*RIB* 741).

An inscription from Birdoswald should not be taken as evidence of the unit being based there. Cohors I Aelia Dacorum seem to have been the garrison of that fort in the Severan era, and so most likely I Thracum was moved north for the Severan campaigns, helping with building work at Birdoswald before being moved back south to their garrison. We have no post-Severan evidence for the unit, and at the end of Roman rule, the *Notitia Dignitatum* says of Bowes: '*Praefectus numeri exploratorum, at Lavatres.*' It is possible that in the intervening century and a half, the Thracian unit has become this 'irregular unit of scouts', of course.

There is some identification difficulty with Thracian cohorts, for there are many of them, including a number of I Thracums. As such, I have not included in this listing one Maximianus at Zama (*AE* 1956, 124), or the *miles* Fuscus Dorilsis at Astorga (IRPLeon 82), since they could feasibly belong to any one of a number of units.

Attested officers and men:
Prefect Lucius Caesius Frontinus at Bowes
Prefect Titus Orbius Priscinus at Bowes
Prefect Claudius Paulus at Tivoli
Centurion Iulius Secundus at Bowes
Trooper Tiberius Claudius Tirintius at Wroxeter

Inscriptions:
Birdoswald (Banna) – granary dedication to Septimius Severus and Caracalla (*RIB* 1909)
Bowes (Lavatris) – building inscription (*RIB* 741)
Bowes (Lavatris) – altar to Fortuna (*RIB* 730)
Bowes (Lavatris) – altar to Vinotonus Silvanus (*RIB* 732)
Bowes (Lavatris) – altar to Vinotonus (*RIB* 733)
Bowes (Lavatris) – unidentified altar (*RIB* 734)
Bowes (Lavatris) – dedication to Septimius Severus, Caracalla and Geta (*RIB* 740)
Bowes (Lavatris) – altar to Vinotonus Silvanus Augustus (*RIB* 3251)
Bowes (Lavatris) – building inscription (*RIB* 3252)
Newcastle (Pons Aelius) – building inscription from the vallum (*RIB* 1323)
Wroxeter (Viroconium) – tombstone of Tiberius Claudius Tirintius (*RIB* 291)

Tombstone of Titus Claudius Tiritius from Wroxeter, now in Shrewsbury Museum. (*Author's collection*)

Cohors II Thracum (aka Cohors Secundae Thracum Veterana Equitata Civium Romanorum Pia Fidelis)

Type: Equitata
Size: Quingenary

Reenactor of Deva Victrix at Park in the Past portraying a centurion of Cohors II Thracum, with a distinctly Thracian look. (*Author's collection*)

Raised from the lands of Thrace (an area that roughly corresponds to eastern Bulgaria and European Turkey), Cohors II Thracum was a unit of 500 mixed infantry and horse. Note at the heading their full and rather grandiose title later in their career. Of their early days we have a tantalizing glimpse. An inscription from Aquileia records the career of one Quintus Etuvius Sextus, a prefect of this unit, and places them in Germany under Nero and/or Vespasian. As such, it is quite possible that they received the '*Pia Fidelis*' honorific during the Civilis Revolt of 69. Cohors II Thracum may be restorable in two diplomas from Lower Germany, dated 80 and 98, and this is likely, given their earlier presence confirmed in the province. This would mean that the unit probably remained stationed in Germany for those thirty or forty years. Indeed, inscriptions on a tile (*AE* 1975, 639f) and a stamped bucket (*AE* 1975, 623) seem to attest the unit still in Lower Germany at Maurik (Mannaricum) perhaps into the Trajanic period.

Of their time in Britain we know nothing concrete until the mid-second century. II Thracum are attested in the Malpas and Brigetio diplomas of AD 103 and 122, which confirms the unit's presence in Britannia. Given that the unit seems to have been in Lower Germany in 98 and in Britain by 103, we can tentatively date their new posting to the beginning of the reign of Trajan. In 98 Trajan was the governor of Lower Germany, and spent the first year of his rule in the province, organizing. As such, he likely shuffled the units around and posted II Thracum to Britain at this time. Of their early service in Britain we can say little. The unnamed Thracian cohort on a tombstone found in Wroxeter is unlikely to be this one, and more probably refers to Cohors I Thracum, given the space available for the numeral. Along with the diploma from Malpas in Cheshire, they do suggest that from AD 99, II Thracum served somewhere in the west or north-west of England. During the occupation of the Antonine Wall, II Thracum are attested on one tombstone at the fort of Mumrills (possibly Pexa?), and it is interesting to note that the occupant of the tomb is recorded as having been one of the tribe of the Brigantes by birth. This lends weight to the unit occupying a fort in Brigantian lands prior to their service on the Antonine Wall. As well as this unit, Ala I Tungrorum are also attested at Mumrills, which suggests that these two were the garrisons of the fort during the two periods of Antonine occupation.

Undated epigraphy places II Thracum at the western shore fort of Moresby (Gabrosentum), and it seems likely that this follows their time at the Antonine Wall, since the *Notitia Dignitatum* places them there at the end of Roman rule in Britain, suggesting constant occupation at Moresby for more than two centuries. The unit had a busy and varied career, earning numerous titles and epithets, and seem to have ended their time with a long and well-deserved rest in garrison. I have included the inscription (*RIB* 805) from Moresby naming a centurion Severinus, despite the lack of a unit confirmation, given its likely Severan dating and the fact that it seems highly likely that II Thracum were based in Moresby at that time.

Attested officers and men:
Prefect Quintus Etuvius Sextus in Lower Germany
Prefect Mamius Nepos at Moresby
Centurion Lucius Severinus
Miles Smertrius, son of Macer at Moresby
Nectovelius, son of Vindex, a Brigantian, at Mumrills

Inscriptions:
Germany – funerary inscription to Etuvius (*ILS* 9090)
Mumrills (Pexa?) – tombstone of Nectovelius (*RIB* 2142)
Moresby (Gabrosentum) – dedication to Jupiter Optimus Maximus (*RIB* 797)
Moresby (Gabrosentum) – building inscription (*RIB* 803)
Moresby (Gabrosentum) – tombstone of Smertrius (*RIB* 804)
Moresby (Gabrosentum) – inscription of Severinus (*RIB* 805)

Cohors I Tungrorum (aka Cohors Primae Tungrorum Milliaria Civium Romanorum)

Type: Infantry
Size: Milliary

The first cohort of Tungrians is one of the most fascinating units of the British *auxilia*, partially because it has one of the most complete records to examine. The very term Tungrian may have been first applied under Augustus, and certainly does not appear in the writings of Caesar, even though he fought through their country and capital, and so we can be confident that any unit drawn from there was not raised until at least the Principate. The Tungrians occupied an area of eastern Belgium centred around Tongeren, and I Tungrorum, along with several other Tungrian cohorts, seem to have been raised prior to AD 69, when two of them (very likely I and II Tungrorum) were sent to Forum Julii during the civil war.

They were likely involved in the Civilis Revolt in Germany the following year, but seem to have remained loyal and were almost certainly not the Tungrian cohort noted to have defected to the rebels. As such, it is extremely likely that the unit (along with its sister cohort, II Tungrorum) were brought from Germany to Britain by Petillius Cerialis in 71. The Tungrians were probably, like their Rhenish Batavian neighbours, skilled in crossing bodies of water in full armour, swimming alongside their horses. It is probable that I Tungrorum was one of the units used by Agricola to invade Anglesey in 78/79, and one of the two Tungrian units present at the Battle of Mons Graupius in Scotland in 83. Tacitus does not give us unit numbers in his work, but the description of the cohorts fits seamlessly with I and II Tungrorum, and so there is no reason to doubt that this is the unit's service record prior to the end of the Flavian era.

At this time, the cohort seems to have been a quingenary unit, with a strength on paper of some 500 men, and from circa AD 85 they are attested at the Stanegate fort of Chesterholm. They seem to have occupied Chesterholm from then to sometime around AD 92 in what is referred to as the Phase I fort. At around that time they moved from there to an unknown location for a short period, before returning. Interestingly, one of the Vindolanda tablets (record 154), dated to early phase II, is a unit strength report and places more than half the unit on active service away from their base in Coria (Corbridge).

At about this time, Cohors VIIII Batavorum moved into Chesterholm and the fort was rebuilt (Phase II) as a much larger affair. It is possible that the fort was enlarged at this time due to the presence of two simultaneous quingenary cohorts (I Tungrorum and VIIII Batavorum). The Batavians transferred out once more around 105, leaving I Tungrorum as the sole garrison, though the unit is seemingly by this time a milliary cohort of 1,000 men, serving under Iulius Verecundus, the same officer who had commanded the quingenary cohort before their departure, as confirmed by the finds of associated tablets in the two phases of the fort. It may be that during this somewhat confused period of occupation (AD 92–105) two of the surviving cohorts of Tungrians following Agricola's wars were combined to create one double-sized unit. The enlarged cohort's presence at Chesterholm is attested in numerous sources, inscriptions, the tablets, and even punched into a spearhead. The cohort seems to have remained there until the beginning of the reign of Hadrian. The spearhead has been dated to AD 122–130, and the tombstone of Centurion Titus Annius records his death in war, which seems to have occurred in the troubles of AD 117 at the beginning of Hadrian's reign. After circa AD 122, only one piece of evidence occurs at Chesterholm, though, with this being a military diploma, it may be that the unit was serving elsewhere and the veteran Amandius simply retired back to familiar Chesterholm.

The name Crispus, relating to the unit in the Vindolanda tablets, is interesting. Though the name is not uncommon, and it may be that the Crispus who wrote to Verecundus in tablet 891 is not the same Crispus in 295, the dates of the two mentions are close enough, falling within around twenty years, that they likely also refer to the same person. If so, then it is possible that Crispus occupied Chesterholm both as part of the initial smaller cohort and then also as part of the milliary cohort upon their return. Crispus is noted as a '*mensor*' (measurer) though without further detail, leaving the interpretation open. He might have been involved with grain collection, or perhaps a military surveyor. The letter in which he is mentioned as such seems to be a request to be granted *immunes* status.

With the construction of Hadrian's Wall, it appears that I Tungrorum was assigned briefly to the fort of Carrawburgh (Brocolitia), for the unit was responsible for a dedication there to the emperor, which dates their presence to AD 122–138. We cannot be certain of their movements around this time. They seem to have left Chesterholm after 122, the year the wall was begun, and the next time they are attested after Carrawburgh is post-140, but during the Hadrianic era, Carrawburgh displays evidence of a dizzying series of units (I Tungrorum, I Aquitanorum, I Cugernorum and possibly I Frisiavonum), all within a few short years. All we can say for sure is that I Tungrorum built something at Carrawburgh during the reign of Hadrian, possibly sharing the original construction of the fort in 130 with I Aquitanorum. There is a suggestion that I Tungrorum, a formerly milliary unit, was once more split around this time into two quingenary cohorts, and that, since a similar unit is attested in

Noricum at the same time this one is mentioned at Carrawburgh, these are the two newly-divided Tungrian units.

With the accession of Antoninus Pius and the creation of the new wall, I Tungrorum moved once again and are next attested in a building dedication to Antoninus at the fort of Castlecary on the Antonine Wall, which places them there between AD 142 and 165. Their activity during the late Antonine era is unknown, and they next appear in the record after 205 in the reign of Septimius Severus.

Severus arrived in Britain in 208 with the intention of conquering Scotland, and the assignment of I Tungrorum to Housesteads may well coincide with this campaign. As part of his plans, Severus reorganized the defences and infrastructure around the two walls, and a new garrison at Housesteads from this time makes sense. I Tungrorum are attested in inscriptions circa 205–208, and Birley dates another inscription of theirs at Housesteads to 213. An altar (*RIB* 1594) that does not name a unit, labels its builders as Germanic recruits of the Tuihanti Frisians, and almost certainly refers to the same Tungrian unit. Interestingly, if these two units at Housesteads are identical, then the unit had by this time acquired the epithet 'of Severus Alexander'. A lengthy stay at the fort is supported by the wealth of epigraphy found there, and indeed, we have every reason to believe that Housesteads remained the unit's base for the duration of

The excellently preserved latrines of Housesteads. (*Author's collection*)

Roman control on the island. Two centuries later, the *Notitia Dignitatum* still places them at Housesteads. This gives us an almost complete unit history over four centuries of activity in Britain.

Attested officers and men:
Prefect Quintus Florius Maternus at Housesteads
Prefect Publius Aelius Modestus at Housesteads
Prefect Quintus Iulius Maximus at Housesteads
Prefect Quintus Iulius [...]sus at Housesteads
Prefect Quintus Verius Superstis at Housesteads
Prefect Iulius Verecundus at Chesterholm
Centurion Decmus at Chesterholm
Centurion Titus Annius at Chesterholm
Medicus Anicius Ingenuus at Housesteads
Beneficarius of the prefect, Hurmius, son of Leubasnus at Housesteads
Calpurnius, heir of Hurmius at Housesteads
Amandius at Chesterholm
Priscinus, probable prefect at Chesterholm
Audax, slave of Iulius Verecundus at Chesterholm
Andangius at Chesterholm
Mensor Crispus at Chesterholm
Crispus at Chesterholm
Vel[...] at Chesterholm

Inscriptions:
Chesterholm (Vindolanda) – strength report (Vindolanda Tablet 154)
Chesterholm (Vindolanda – letter (Vindolanda Tablet 295)
Chesterholm (Vindolanda) – strength report (Vindolanda Tablet 857)
Chesterholm (Vindolanda) – letter fragment (Vindolanda Tablet 893)
Chesterholm (Vindolanda) – military diploma (*RIB* II, 2401.9)
Chesterholm (Vindolanda) – spearhead inscription 'TVNG'
Chesterholm (Vindolanda) – dedication to Ahvardva
Chesterholm (Vindolanda) – tombstone of Titus Annius (*RIB* 3364)

Tombstone of the medic Anicius Ingenuus from Housesteads, now in the Hancock Museum. (*Author's collection*)

Carrawburgh (Brocolitia) – dedication to Hadrian (*RIB* 3317)
Housesteads (Vercovicium) – altar to Silvanus Cocidius (*RIB* 1578)
Housesteads (Vercovicium) – altar to Hercules (*RIB* 1580)
Housesteads (Vercovicium) – altar to Mars, the two Alaisiagae and the divinity of the emperors (*RIB* 1594)
Housesteads (Vercovicium) – altar to Jupiter Optimus Maximus and the divinity of the emperors (*RIB* 1584)
Housesteads (Vercovicium) – altar to Jupiter Optimus Maximus and the divinity of the emperors (*RIB* 1585)
Housesteads (Vercovicium) – altar to Jupiter Optimus Maximus and the divinity of the emperors (*RIB* 1586)
Housesteads (Vercovicium) – altar to Jupiter Optimus Maximus and the divinity of the emperors (*RIB* 1588)
Housesteads (Vercovicium) – altar to Mars (*RIB* 1591)
Housesteads (Vercovicium) – altar to the Mother Goddesses (*RIB* 1598)
Housesteads (Vercovicium) – dedication to Septimius Severus, Caracalla, and Geta (*RIB* 1612)
Housesteads (Vercovicium) – dedication by instruction of Clarian Apollo (*RIB* 1579)
Housesteads (Vercovicium) – tombstone of Anicius Ingenuus (*RIB* 1618)
Housesteads (Vercovicium) – tombstone of Hurmius (*RIB* 1619)
Housesteads (Vercovicium) – building stone (*RIB* 3326)
Housesteads (Vercovicium) – possible inscription (*AE* 1967, 260)
Castlecary – building dedication to Antoninus Pius (*RIB* 2155)
For the split between Carrawburgh and Noricum (*AE* 1967, 249)

Cohors II Tungrorum (aka Cohors Secundae Tungrorum Civium Liberorum Milliaria equitata)

Type: Equitata
Size: Milliary

Like its sister cohort, I Tungrorum, this unit was a 1,000-strong force of mixed cavalry and infantry, formed from the Germanic tribe of the Tungri in the area of Tongeren in modern Belgium. The warriors of the region were prized as cavalry from early days, and Julius Caesar relied heavily upon these tribes during his Gallic and Civil Wars. The emperor's German Guard may even have been drawn from among the Tungri. Cohors II Tungrorum seems to have been commanded by a man of prefect rank throughout its existence, an unusual situation for a milliary unit.

II Tungrorum's precise origin is unknown, although it was likely raised during the Julio-Claudian era. It is possible that the cohort was one of the Tungrian units (two infantry and four cavalry) that Fabius Valens sent to garrison Forum Julii during the civil war of AD 69. It is highly likely that it was involved in the revolt of Civilis in Germany in 70 and if the 'C L' included in the unit's name on no less than six sources can be expanded as '*Coram Laudata*' ('Publicly Praised'), the epithet may have been granted for the unit's activity during this revolt, when a Tungrian unit, which does not appear to have been this one, defected to the enemy. An alternative expansion of the 'C L' could be '*Civium Liberorum*'. As '*Civium Romanorum*' ('Roman citizens') appears on many unit names, '*Civium Liberorum*' would mean free citizens, a notable thing for an auxiliary unit, and which might explain the lack of evidence for the unit in diplomas. If they were already free, they would not need such a thing on discharge.

The unit was then likely brought across to Britannia by Petillius Cerialis in AD 71 alongside another Tungrian unit and Legio II Adiutrix. Its locations and activity in its early days on the island are unknown, although when Agricola governed the island, it seems likely the unit was involved in his varied campaigns. In AD 77/78 they may have been based somewhere in the Midlands or in North Wales, for Agricola is noted as having gathered a force between landing in the south of England and invading North Wales and Anglesey, and among that force were auxiliaries who were capable of swimming the Menai Strait in armour with their horses. This is a description often applied to Germanic cavalry, and may include the Tungrians. Once again, Agricola is noted to have fielded two Tungrian cohorts at the Battle of Mons Graupius in 83, which lends a heavy suggestion to the presence of this unit. As such, then, II Tungrorum likely served all across North Wales, the North of England, and then Scotland between 77 and 83.

Where they were stationed from this time until the first confirmed epigraphy we cannot be sure, though they are likely to have continued to be based in Scotland and the North of England, for they are next found in the reign of Antoninus Pius at the fort of Cramond on the outskirts of Edinburgh. Cramond was not part of the Antonine Wall *per se*, but seems to have been connected to it, perhaps in the form of a harbour/supply base, in the same manner as South Shields for Hadrian's Wall. It is entirely possible that there was an earlier fort on the site from the time of Agricola, but it seems that Cohors II Tungrorum were involved in the building of a fort here circa 140, under the watchful eye of a centurion from Legio XX Valeria Victrix. They may have been the garrison here for a time, or were perhaps just part of the Antonine building force in the region, as suggested by their being overseen be a legion. Cramond seems to have had four phases of occupation, two of them Antonine, and it seems reasonable to suggest that II Tungrorum were the first, with I Cugernorum during the second Antonine phase, V Gallorum during the Severan campaigns, and an unknown unit thereafter.

Interestingly, concurrent with their occupation of Cramond, in 146 and 153 they are also attested by the settlement of veterans in Raetia at no less than three different forts: Straubing (Sorviodurum), Manching (Vallatum) and Gnotzheim (Mediana). It is highly likely that vexillations of these prized German cavalry were brought to the mountains of mainland Europe during the reign of Antoninus Pius for specific campaigns and then, once the veterans were settled, sent back to Britain. They are certainly absent from Raetia by 157. It is likely that II Tungrorum remained the garrison of Cramond for almost two decades before being moved, later in the reign of Antoninus Pius, to Birrens circa AD 156, when they are clearly attested. There is plentiful evidence of their presence at Birrens, one of the advance forts of Hadrian's Wall, before they are moved on once more, and, given the dates, it is likely the units from Cramond and Raetia were recombined at Birrens.

Their apparently final move took place somewhere between the end of the reign of Antoninus Pius in AD 161 and the time of Gordian III in 241. They then appear repeatedly in inscriptions at the fort of Castlesteads on Hadrian's Wall, just 20 miles from their previous home at Birrens. It seems likely that the unit remained at Camboglanna from 241 at the latest until the end of Roman

Altar from Birrens in Dumfries Museum. (*Author's collection*)

Reenactors portraying Tungrian auxiliaries. (*Author's collection*)

Britain. The *Notitia Dignitatum*, a list of Britannia's military at the dawn of the fifth century, suspiciously omits both Banna and II Tungrorum from its list, as well as missing the C from Camboglanna. As such it seems highly likely that there is a lacuna in the text of the *Notitia Dignitatum* that could be reconstructed as: 'Tribunis cohortis primae Aeliae Dacorum, [*Banna Tribunis cohortis secundae Tungrorum, C*]amboglanna'. This seems convincing enough concerning the existence of the unit in that home fort until the very end.

Attested officers and men:
Prefect Albius Severus at Castlesteads
Prefect Aurelius Optatus at Castlesteads
Prefect Tiberius Claudius Claudianus at Castlesteads, AD 241
Prefect Publius Campanius Italicus at Birrens
Prefect Gaius Silvius Auspex at Birrens
A centurion of Legio XX VV, possibly a weapons instructor
Centurion Vic[…] Severus at Castlesteads
Centurion Messius Opsequens at Castlesteads
Centurion Publius Aelius Martinus, AD 241
Centurion Afutianus, son of Bassus
Celer, freedman of Publius Campanius Italicus
Frumentius, *Miles*
Flavia Baetica, wife of Afutianus

Inscriptions:
Birrens (Blatobulgium) – altar to the discipline of the emperor (*RIB* 2092)
Birrens (Blatobulgium) – Mars and the victory of the emperor (*RIB* 2100)
Birrens (Blatobulgium) – unidentified Altar (*RIB* 2101)
Birrens (Blatobulgium) – altar to Minerva (*RIB* 2104)
Birrens (Blatobulgium) – altar to Ricagambeda (*RIB* 2107)
Birrens (Blatobulgium) – altar to Viradecthis (*RIB* 2108)
Birrens (Blatobulgium) – altar to all gods and goddesses (*RIB* 2109)
Birrens (Blatobulgium) – dedication to Fortuna (*RIB* 2094)
Birrens (Blatobulgium) – dedication to Antoninus Pius (*RIB* 2110)
Birrens (Blatobulgium) – funerary inscription for Afutianus (*RIB* 2115)
Cramond – altar to the Alatervae and the Mother Goddesses of the parade ground (*RIB* 2135)
Castlesteads (Camboglanna) – altar to Jupiter Optimus Maximus (*RIB* 1981)
Castlesteads (Camboglanna) – altar to Jupiter Optimus Maximus (*RIB* 1982)
Castlesteads (Camboglanna) – altar to Jupiter Optimus Maximus and the divinity of the emperor (*RIB* 1983)
Castlesteads (Camboglanna) – unidentified dedication slab (*RIB* 1999)

Cohors I Vangionum (aka Cohors Primae Vangionum Milliaria Equitata)

Type: Equitata
Size: Milliary

Like many of Britain's auxiliary units, Cohors I Vangionum was drawn from a Germanic tribe, in this case the Vangiones, who occupied an area covering both banks of the Rhine in the area of Mainz. I Vangionum may have been created during the Julio-Claudian era, though we have no knowledge of their activity until AD 67, when they are mentioned in one of the Bloomsberg tablets. Where they were based in Britain at this time we cannot say, but we can be confident that they were on the island by then, and possibly even during the island's invasion in 43. The Malpas diploma from Cheshire dated to 103 suggests a garrison location somewhere in Wales or the West Midlands under Trajan.

Their first possibly attested garrison is still a little uncertain. An altar found at Clifton in Cumbria names a centurion from a unit that may have been based at the nearby fort of Brocavum at Brougham. The inscription is incomplete but it labels him as a man of 'Cohors I V', which could be either Cohors I Vangionum or Cohors I Vardullorum, although the latter is often preceded by the appellation 'Fida', so it is more likely to represent this unit. If the identification is correct, then this could suggest Brougham as the base of the cohort during the Hadrianic era. Certainly the Brougham fort is of a size appropriate for a milliary unit of 1,000 men and so would suit the Vangiones. The finding in 2014 of an altar at Papcastle, however, offers an alternative. The altar names COH IV again, and without '*Fida*', again likely names the Vangiones, and Papcastle could also house such a garrison. It is, of course, possible that the unit occupied both these forts at different times. That a diploma suggests a member of the unit retired in the Ravenglass area in AD 158 provides another Cumbrian link that might support a base in the region for a time.

The next place we find I Vangionum, and the place where they likely spent the rest of Roman rule in Britain, was at Risingham, one of the advance forts of Hadrian's Wall on Dere Street. We cannot be certain when they took command of the fort, but it seems to have been late in the reign of Septimius Severus. When Severus began his campaign against Scotland in AD 208 he rearranged the forts, supply lines, and garrisons of the walls of both Hadrian and Antoninus. As such it seems likely that I Vangionum was brought to Risingham as part of his reorganization, securing the military road into the North. Epigraphy supports the notion that the unit took control of a rundown fort and rebuilt it. One inscription found at the south gate of the fort confirms that the structure had fallen in and was rebuilt from the ground up by the unit between AD 205 and

208. A second inscription, dated 213, not long after the death of the emperor, also confirmed another unknown structure rebuilt by the unit, suggesting that the work was ongoing after the end of the war and the reoccupation of Hadrian's Wall under Caracalla. In addition to I Vangionum, these inscriptions mention a unit of scouts, as well as a unit of Raetian spearmen, based here alongside the cohort and helping with the rebuild.

Inscriptions (at least two altars) place the unit at Benwell on Hadrian's Wall at some unknown date, but since the Benwell fort could realistically accommodate only a quingenary unit of 500 men, it seems likely that this fort was occupied by only half the cohort, with the other half in a second location. It may even be that only a vexillation of the unit were based at Benwell while the rest remained in garrison at Risingham, or vice versa, the distance between the two being a manageable 30 miles. Indeed, the two inscriptions from Benwell label the unit commander as a prefect, rather than the usual tribune, and this may be an indication that the unit was split for a time into two quingenary subunits, commanded by prefects. This would neatly explain the command ranks we find in epigraphy and their ability to occupy smaller forts.

Risingham is not mentioned in the *Notitia Dignitatum*, and the fort was abandoned long before the turn of the fifth century. There is no sign of a change of garrison after the arrival of I Vangionum, and so they probably occupied the fort until its abandonment, whenever that occurred. What happened after that we cannot be certain, but there is a tantalizing possibility raised by an altar found at Gloster Hill near Amble in coastal Northumberland. An unidentified 'Cohors I' raised the altar, and the nearest fort is that of Learchild (Alauna) on which no excavation has occurred and whose garrison is unknown. Alauna is certainly large enough to accommodate the cohort, and so it is just possible that this little-known fort housed I Vangionum in the post-Severan period.

A tombstone at Chesters does not suggest that the cohort was based there, but rather that a veteran retired there. It is worth noting that Chesters is only 10 miles from Risingham, 20 from Benwell, and 30 from Learchild. Another (undated) tombstone from Colchester again does not suggest the unit's presence so much as that of a retiree. This tombstone reads only 'I VA[…]', and might be a reference to another unit.

Attested officers and men:
Tribune Iulius Paullus at Risingham
Tribune Lucius Aemilius Salvianus at Risingham, AD 205–208
Tribune Aemilius Aemilianus at Risingham
Tribune Iulius Victor at Risingham
Tribune Marcus Peregrinius Super at Risingham

Tribune Fabius Honoratus
Prefect Cassianus at Risingham
Prefect of cavalry Tineius Longus?
Centurion Subrius Apollinaris of the First Cohort at Brougham
Decimus Iulius Candidus, son of Quintus at Benwell
Aurelia Eglectiane, wife of Fabius Honoratus
Fabia Honorata, daughter of Fabius Honoratus

Inscriptions:
Amble – altar to the Goddesses of the Parade Ground (*RIB* 1206)
Benwell – altar to Antenociticus (*RIB* 1328)
Benwell – altar to Anociticus (*RIB* 1329)
Benwell – tombstone to Decimus Iulius Candidus (*RIB* 1350)
Chesters – tombstone to Fabia Honorata (*RIB* 1482)
Clifton (Cumbria) – altar to Jupiter Optimus Maximus and the *genius loci* (*RIB* 792)
Colchester (Camulodunum) – tombstone (*RIB* 205)
Papcastle (Derventio) – altar to Mars (*BRIT.* 46.6)
Risingham (Habitancum) – altar to Hercules (*RIB* 1213)
Risingham (Habitancum) – altar to Hercules (*RIB* 1215)
Risingham (Habitancum) – altar to Jupiter Optimus Maximus (*RIB* 1216)
Risingham (Habitancum) – altar to Jupiter Optimus Maximus (*RIB* 1217)
Risingham (Habitancum) – altar to the Mother Goddesses from Overseas (*RIB* 1224)
Risingham (Habitancum) – unidentified altar (*RIB* 1230)

Temple of Anteniciticus at Benwell. (*Author's collection*)

Risingham (Habitancum) – unidentified altar (*RIB* 1231)
Risingham (Habitancum) – dedication slab on rebuilding (*RIB* 1234)
Risingham (Habitancum) – building dedication to Caracalla and Julia Domna (*RIB* 1235)
Risingham (Habitancum) – unidentified building inscription (*RIB* 1241)
Risingham (Habitancum) – unidentified building inscription (*RIB* 1243)

Cohors I Vardullorum (aka Cohors Primae Fida Vardullorum Civium Romanorum Equitata Milliaria)

Type: Equitata
Size: Milliary

A part-mounted, 1,000-strong unit founded among the northern Iberian Vardulli people in the Basque region, this unit would have included prized cavalry. The northern Hispanic riders' skill and tactics had been so iconic since at least the first century BC that their famous cavalry manoeuvre 'The Cantabrian Circle' was quoted in Roman military manuals.

Though we know where they were founded, the date is much less certain. Cohors I Vardullorum attainted the epithet '*Fida*' ('Faithful') early in their career, and this may be a reference to involvement in the civil war of AD 69 on the side of Vespasian, for the armies of Gaul and Hispania largely came over to the Flavian banner without a fight. Whatever the case, we have no confirmable details of their existence prior to AD 98, when they first appear in the records of Britannia. Their early garrisons and movements on the island are uncertain, but they are attested in military diplomas from AD 98 through to 159.

While it is possible that I Vardullorum were based early on in Brougham (Brocavum), the inscription that allows for this (which lacks the word '*Fida*') more realistically points to Cohors I Vangionum, and so their presence seems unlikely. Our first solid hint as to their occupation comes from the Antonine Wall. Castlecary fort was occupied from circa AD 142 to 165 and likely sporadically afterwards, including during the campaigns of Septimius Severus. A cohort of Tungrians were seemingly based at the fort at some point during that time, but it is possible that Cohors I Vardullorum was also here during that period, for an altar was set up at the fort by their prefect. Indeed, it seems there were two distinct occupations of the Antonine Wall, during the reign of the eponymous emperor, and early in the reign of his successor, and we may be looking at two units spread over these two periods. The Vardulli were perhaps the first unit based at the fort, in the Antonine period, with the Tungrians moved in to replace them under the next emperor. That, while at Castlecary, the unit was commanded by a prefect and not a tribune suggests the possibility that only part of the unit was based there.

Where they moved when the wall was abandoned in AD 165 and Roman forces were withdrawn back behind Hadrian's Wall we cannot say for sure. The earliest potential evidence comes from a stone found in Hexham (*RIB* 1128), which references the 'emperors', and so cannot predate the joint rule of Marcus Aurelius and Lucius Verus from AD 161. This might suggest a post-Antonine position somewhere in the area of Corbridge. It is more likely, though, that this stone comes from the nearby fort of Longovicium at Lanchester, for that is the next known garrison site of the cohort, only a decade or so later, and lies only 20 miles away.

Lanchester, one of the quieter forts on Dere Street, is confirmed as their base around AD 175–178 by two altars. A centurion of a 'COH V' attested in building work at nearby Ebchester likely references a centurion from the fifth cohort of one of the legions. It may, however, be that the I Vardullorum helped with work on their neighbouring fort, only 6 or 7 miles away.

Cohors I Vardullorum appear to have been moved to the outpost fort of High Rochester before AD 213. It seems likely that this was part of Caracalla's reorganization of the frontier after his father's death and the end of the Scottish campaign. Roman forces were withdrawn to Hadrian's Wall, and so High Rochester would suddenly become important once more. It is interesting to note that the unit they relaced, I Lingonum, moved to Lanchester, the two units essentially swapping forts during the Severan campaigns. They continue to be attested at High Rochester into the mid-third century, and two of the inscriptions from that fort commemorate the construction of Artillery Platforms outside the walls. The fort of Bremenium does not appear in the *Notitia Dignitatum*, and so the fort seems to have been abandoned before the fifth century.

We have no record of the activity of I Vardullorum after 241. They may have occupied Learchild (Alauna), just 20 miles from their previous garrison of High Rochester, based upon *RIB* 1206 which was found at Gloster Hill, near Amble on the Northumberland coast, and which references a 'Cohors I', and that they are the most local unit adds weight to this suggestion. Two altars were raised by one of the unit's commanders somewhere in the region of Jedburgh, and possibly originate from the nearby fort at Cappuck, or perhaps indicate long range patrols from High Rochester, less than 20 miles away. An altar set up by the unit in Milecastle 19 on Hadrian's Wall may have been moved from its original position, or may designate a temporary posting for a vexillation from either Lanchester or High Rochester. We have no definitive record of the unit after its third-century occupation.

Interestingly, the cohort may have, for a time in the second century, been split, with a vexillation sent elsewhere, for the word 'milliary' is missing from their name on diplomas of the period, and this would account for their being commanded for a time by a prefect rather than a tribune.

Attested officers and men:
Tribune Titus Flavius Titianus in Lanchester, AD 175–178
Tribune Titus Licinius Valerianus at High Rochester
Tribune Cassius Sabinianus at High Rochester, AD 238–241
Tribune Lucius Caecilius Optatus at High Rochester, AD 213
Tribune Publius Aelius Erasinus at High Rochester
Tribune Aurelius Quintus at High Rochester
Tribune Gaius Quintius Severus, of the Camilian voting-tribe from Ravenna
Prefect Trebius Verus at Castlecary
Prefect Marcus Sitius Constans, from Cirta in Africa
Prefect Publius Aelius Atticus AD at Lanchester, AD 175–178
Publius Calpurnius Victor, post 161 due to 'emperors' reference
Centurion Martialis, Fifth Cohort in Ebchester
Centurion Subrius Apollinaris of the First Cohort at Brougham
Julia Lucilla, wife of Rufinus
Publius D… V…

Inscriptions:
Amble – altar to the Goddesses of the Parade Ground (*RIB* 1206)
Castlecary – altar (*RIB* 2149)
Clifton (Cumbria) – altar to Jupiter Optimus Maximus and the *genius loci* (*RIB* 792)
Ebchester (Vindomora) – centurial stone of Martialis (*RIB* 1110)
Hexham (Corbridge?) – dedication to the discipline of the emperors (*RIB* 1128)
High Rochester (Bremenium) – altar to the *genius* of our lord and the standards of the First Cohort (*RIB* 1262)
High Rochester (Bremenium) – altar to the *genius* and the standards of the First Cohort (*RIB* 1263)
High Rochester (Bremenium) – altar to Minerva and the *genius* of the collegium (*RIB* 1268)
High Rochester (Bremenium) – dedication to Sol Invictus (*RIB* 1272)
High Rochester (Bremenium) – building dedication (*RIB* 1279)
High Rochester (Bremenium) – building inscription (*RIB* 1285)
High Rochester (Bremenium) – artillery platform building dedication (*RIB* 1280)
High Rochester (Bremenium) – artillery platform building dedication (*RIB* 1281)

Dedication to Alexander Severus from High Rochester, now in the Hancock Museum. (*Author's collection*)

High Rochester (Bremenium) – tombstone to Rufinus (*RIB* 1288)
Jedburgh – altar to Jupiter Optimus Maximus (*RIB* 2117)
Jedburgh – altar to Jupiter Optimus Maximus (*RIB* 2118)
Lanchester (Longovicium) – altar to Aesculapius (*RIB* 1072)
Lanchester (Longovicium) – altar to the emperor's fortune (*RIB* 1073)
Lanchester (Longovicium) – altar to Jupiter Optimus Maximus (*RIB* 1076)
Milecastle 19 – altar to the Mother Goddesses (*RIB* 1421)

COHORTS OF UNCERTAIN STATUS OR SCARCE DATA

Cohors I Afrorum Civium Romanorum

Confirmed in Britain on the Brigetio diploma of AD 122 and possibly on the AD 124 diploma, we have no other evidence of the unit.

Cohors I Alpinorum Peditata

Raised in the Alpine region to the north of Italy, this cohort's date of origin is unknown, though it may have been an early, Augustan creation. There are two known units by this name that are well attested, and it is possible that the one in Britain is a third, otherwise unknown one, as the British service fits poorly into either of the two units' records. We are perhaps best suggesting that this is a totally different Cohors I Alpinorum, raised under Augustus and unknown other than on the 103 diploma in Britain.

Cohors I Asturum

This cohort is only mentioned in one source. The *Notitia Dignitatum* tells us of a '*Tribunus cohortis primae Asturum, at Aesica*.' There is some debate over this. At first glance it seems that I Asturum is a transcription error in the text, and that it should refer to II Asturum, who are already well attested at that site. However, an inscription (*CIL* VIII, 20736) from Souk-el-Ghozlane in Algeria places its subject (Prefect Quintus Gargilius Martialis) as commander of '*cohortis I Asturum provinciae Brittanniae*', which confirms that unit's presence in the province. As such it is at least worth considering the possibility that in the third or fourth century, II Asturum were replaced at Great Chesters by their sister cohort, I Asturum, while they themselves moved on to some other unknown garrison not mentioned in the *Notitia Dignitatum*. Of course, given the changes in the military over the fourth and fifth centuries, it is not impossible that the two Asturian cohorts, both seriously reduced, were combined, taking the numeral of the older unit, and it is therefore both cohorts in effect that occupy Great Chesters.

Cohors IV Asturum Equitata

The existence of this unit in Britain is likely erroneous. Their presence has been suggested by the inscription *RIB* 2032 in Carlisle Castle, which carried the text 'COH IIII ASTA' though this is more likely to be translatable as Cohortis IIII Centuria Hastati, a building inscription of the *hastatus prior*, a centurion of the fourth cohort. It is likely a building inscription of one of the three legions involved in the construction of the wall and should not be taken as evidence of this cohort's presence in Britain. It is worth noting, though, through simple interest, that we have evidence empire-wide for cohorts I, II, III, V and VI Asturum, but no IV, though that does not change the evidence.

Cohors III Batavorum

Attested only on Vindolanda tablets, there is no other record of this unit in Britain. It may be attested in the second century in Raetia or Pannonia.

Cohors VIII Batavorum

Otherwise unattested in Britain, this cohort seems to have been the garrison of Chesterholm (Vindolanda) around AD 102, based on the evidence of the Vindolanda tablets. One of the more famous of the tablets, a birthday invitation from Claudia Severa, is addressed to Sulpicia Lepidina, wife of Flavius Cerialis, the commander of Cohors VIII Batavorum. Unless this was a mistake, this places the unit there at the time of the letter. The only other evidence for the unit is based on military diplomas that place them in Raetia in the Hadrianic era. They were likely another unit that was raised in the aftermath of the Civilis Revolt, probably served under Cerialis and Agricola in Britain, remaining in garrison at Chesterholm through the Trajanic era, and were then transferred to Raetia, following which we lose track of them.

Cohors IX Batavorum

Stamped leatherwork at Chesterholm (Vindolanda) strongly suggest that IX Batavorum were in garrison there at some point, athough they are attested in Raetia by AD 116.

Cohors I Bracaraugustanorum

The existence of this unit in Britain is disputed. The one piece of evidence we have does not name the unit directly. *RIB* 649 found in York is an altar to Jupiter raised by one Publius Aelius Marcianus, the prefect of an unnamed cohort. Birley has linked this name to a Publius Aelius Marcianus named on a career inscription in Mauretania (*ILS* 2738), who was the prefect of Cohors I Augusta Bracarum, and through this suggests their presence in Britain. While this is possible, it is far from proof. That inscription lists a broad career with numerous units. It is highly unlikely that the Publius Aelius Marcianus in York is the same one as on the Mauretania inscription, and so this unit can safely be overlooked.

Cohors I Celtiberorum

Raised in the Tarraconensis region of Spain, and taking its name from the pre-Roman people of the region, the generally accepted date of their founding is the late first century. There is no evidence of a garrison outside Britain prior to their inclusion on the British diploma of AD 105, and it is entirely probably that they were raised in Iberia and shipped directly to Britain. A question has been raised over diplomas from Mauretania Tingitana, all dated AD 109, which also reference a Cohors I Celtiberorum. That the unit was transferred there and back is unlikely, and the discovery of epigraphy at Castromao in Spain, dating to AD 132 when our unit is in Britain according to diplomas, makes it almost certain that there were two entirely different units with the same name and number, one serving in Morocco and the other in Britain. As such, I have not included any reference to Spanish and African sources, and concentrated solely upon British evidence.

Sadly, said British evidence is almost nil. Diplomas confirm the unit's presence between AD 105 and 178, but no inscriptions have yet been unearthed that reference the unit. The only direct evidence we have of their presence beyond diplomas is a series of tiles stamps from the fort of Caersws in Wales, which bear the legend 'C I C F', expanded to read '*Cohors I Celtiberorum Figlina*' ('the pottery works of the unit') or possibly simply 'Cohors I Celtiberorum *fecit*' ('made this').

Cohors I Cornoviorum

The existence of the unit in Britain is questionable, and its history is a blank sheet until the end. The only evidence for the cohort comes from the *Notitia Dignitatum*, which places them at the fort at Pons Aelius (Newcastle) on Hadrian's Wall. They are notable for being the only unit stationed in Britain that was raised in the same province. The Cornovii were centred around Wroxeter in Shropshire, and so the unit must have originated in that area. Roman imperial military administration avoided garrisoning troops in the region in which they were raised, though such rules may have long been abandoned by the time of the *Notitia*. For the possibility of a mistake in the unit name, see Cohors I Cugernorum.

Cohors II Delmatarum

Little is known for certain about Cohors II Delmatarum. They were raised on the Dalmatian coast at some unknown date and sent to Britain at an equally uncertain time. They may have been involved in the invasion in AD 43, though this is merely conjecture. The diploma of AD 105 that is often taken as the first evidence of their presence on the island names only '[…]MATA' which could equally refer to IV Delmatarum. The AD 122 diploma confirms their presence by that year. There is a suggestion that the altar to Virgo Caelestis at Carvoran (*RIB* 1791) belongs to this unit, which would make them the garrison there from the late Severan era onwards, though it is more reasonable to associate that altar with I Hamiorum. Indeed, the only epigraphic evidence to date that names the unit at all is the altar to Veteris (*RIB* 1795) found at Thirlwall. Thus the only things we can say for certain about their service record is that they were in Britain by AD 122, and by the turn of the fifth century they were based at Carvoran, as confirmed by the *Notitia Dignitatum*.

Cohors IV Frisiavonum

It is somewhat doubtful that this cohort existed at all. The only evidence comes from an inscription from Bowes now lost, in which Camden noted the line 'COH IIII F', which he expanded to 'Frisiavonum'. Birley, suggests that the letter was a B and not an F, and therefore refers to Cohors IIII Breucorum, a unit known to have been in Britain, at both Castleshaw and Ebchester.

Cohors VI Gallorum

The only evidence for this unit's existence is an inscription from Rome (*CIL* VI, 1449), and it seems likely that the unit number is a mistake in carving, and that it should reference V Gallorum, to whom Spaul assigns the prefect.

Cohors III Lingonum

This *equitata* unit certainly existed, and is referenced on no less than four diplomas, placing it in Britain between AD 98 and 178. Unfortunately, the only other potential evidence for the cohort in Britain is a lead seal found at Aldborough (*BRIT.* 52.31), which may denote either this unit or possibly IIII Lingonum. The unit is named on a dedicatory inscription in Acqualagna and possibly others from Embres-et-Castelmaure and Tarragona, though they shed no further light on the unit's history.

Cohors I Menapiorum Nautarum

The Menapii were a coastal Belgic tribe, their territory extending up to the Rhine (see also I Morinorum, their neighbours). The tribe's familiarity with the sea shows throughout history, a late Roman unit drawn from them noted as based on the Rhine, the breakaway emperor and fleet commander Carausius of their stock, and the one Menapi unit known in Britain referred to in one of three diplomas as '*nautarum*'. Though we have no inscriptions to mark their presence, diploma evidence places them in the province between at least AD 122 and 135. The only potential epigraphic evidence we have of the unit at all is a leather scrap from Chesterholm stamped with 'C I M', which, if it refers to a unit and not an individual, could be either I Menapiorum or I Morinorum. There is a suggestion that the unit, being sailors, were more mobile than most cohorts, and spent most of their time in leather tents, rather than stone structures.

Cohors I Morinorum

The Morini were a coastal Belgic tribe, based around what is now the Pas de Calais in France (see also I Menapiorum, their neighbours). Their origin date is unknown, but they are attested in Britain on diplomas between AD 103 and 178. They are named on an inscription in Solin, Croatia, which identifies one of their prefects, Quintus Servilius Pacuvianus, and probably ('ORINOR') on another from Grumentum, naming Prefect Gaius Mulvius Ofillius. Despite their lack of epigraphy in Britain, which may point to a mobile, nautical role, (see I Menapiorum and also for Chesterholm evidence) they were active in the province for at least three centuries, still present according to the fifth century *Notitia Dignitatum*, which places them at Glannibanta. (Glannoventa, commonly identified with Ravenglass in Cumbria, notably a coastal station for a coastal people.)

Cohors I Nauticarum

The only evidence for this unit comes from a diploma of AD 135 which, among the units listed in Britain, contains the text [...]N[...]NAUT[...]. There is a strong suggestion that this cohort is the same as Cohors I Aelia Classica, raised from the fleet and connected with Ravenglass. Since Aelia Classica appears on the diplomas of 145 and 158 but not on the 135 one, this is the most acceptable explanation. See also III Nerviorum.

Cohors III Nerviorum

Though we know this unit served in Britain from three diplomas dated AD 122–135, we have no concrete epigraphic evidence of their stay. The Chesterholm inscription (*RIB* 1691) sometimes attributed to them cannot now be read convincingly, and even upon its early discovery, Wallis could only make out Nerviorum without a number. Similarly, the fragment from Maryport (*AE* 1967, 257) names only a 'COH III N[...]'. This could feasibly be our unit, for with only one exception the only units that begin with N in Britain were Nervian. However, that one exception (I Nauticarum) may have been based at Ravenglass just 25 miles down the coast. Cohors III Nerviorum is stated in the *Notitia Dignitatum* as being stationed at Ailione, which is an unknown station, most commonly believed to be the same as Calunium and applied to Lancaster. It is possible that Ailione is a corruption of Alauna, the name of Maryport, which would support the case for their presence there, though that would put it out of order on the listing, and so is unlikely. Simply the evidence for the unit's history at the time of writing is sketchy and confused and tells us nothing with any certainty other than that they were in Britain for at least thirteen years.

Cohors IV Nerviorum

This cohort, one of several Nervian units based in Britain, is only known from one confirmed source, the diploma of AD 135. It may be also noted on an undated diploma found in Tarazona, Spain, though the IIII NERVI mentioned there is disputed, with some scholars preferring an interpretation of I ET III NERVI, which would reference different units. Regardless, at least one diploma places them in Britain, and the *Notitia Dignitatum* confirms that a unit named Numeri Nerviorum Dictensium was in place at Dicti (possibly Whitby) at the end of Roman Britain, which may well denote a unit reduced in number (*numerus*) by that time, especially as their commander was a prefect, while tribunes are noted in command of the other Nervian units in the *Notitia*. Given that I, II, III and VI Nerviorum are all attested in Britain, it is highly probable that both IV and V Nerviorum were based here also, further supporting the unit's presence.

Cohors I Pannoniorum

The existence of this unit in Britain is meagre, yet undeniable. The only confirmed evidence we have is an inscription from Aesernia (*CIL* IX, 2649), which names one Publius Septimius Paterculus as a commander of Cohors I Pannoniorum in Britannia. The inscription makes it likely that Paterculus commanded the unit in the Hadrianic or early Antonine period. The only epigraphic evidence in Britain is a doubtful interpretation of *RIB* 1667, found on Hadrian's Wall at Cawfields Milecastle, the tombstone of one Dagvalda, a soldier of a Pannonian unit. Unfortunately, due to the missing text, the only things that can be made out clearly are 'M.' (*miles*) and 'PAN', indicating a Pannonian unit. As such, it is more likely that the stone references Cohors II Pannoniorum.

Cohors V Pannoniorum

We cannot be certain that this unit served in Britain at all. The only potential evidence for them comes from a lead seal found among many others at Brough-under-Stainmore, which bears among other markings, the main legend 'C V P'. That this refers to 'Cohors V Pannoniorum' cannot be said for sure, but it also cannot be ruled out, given the existence of I and II Pannoniorum here.

Cohors V Raetorum

We know V Raetorum, raised from among the tribes of Raetia, in the northern Alps, served in Britain solely thanks to the diploma of AD 122. An inscription from Bejaia in Mauretania Caesariensis names one Sextus Cornelius Dexter as a prefect of the cohort, but without any indication that this was in Britain. Two inscriptions have been tentatively linked with the unit. The first is a building inscription from Great Chesters (*RIB* 1737) that names a 'RAETORUM' cohort, though the numeral is illegible. It could conceivably be V, or VI, or any other number ending in a I. The other is an altar from Coventina's Well at Carrawburgh (*RIB* 1529), which Birley suggested might refer to this unit, though the text is barely legible and seems not to match up closely enough for comfort. As such, all we can say for certain about V Raetorum is that they were in Britain in AD 122 and that Cornelius Dexter commanded them at some point. For other possible activity linked with this unit, see the irregular unit Vexillatio Gaesatorum Raetorum.

Cohors VI Raetorum

Cohors VI Raetorum's history in Britain is troublesome. A Cohors VI seems to have served in Moesia in the first century and then been transferred to Cappadocia, and Cohors VI Valeria Raetorum is attested at the end of the

Lorica segmentata from Great Chesters in the Hancock Museum. (*Author's collection*)

fourth century in Raetia, and yet the evidence does suggest that they served for a time in Britain. They have been tentatively linked with the building inscription of Great Chesters (*RIB* 1737), though that could just as easily refer to V Raetorum, but the existence of no less than eight lead seals among the large number found at Brough-under-Stainmore, all bearing the inscription CVIR, seems to confirm beyond doubt their service on the island. See also Vexillatio Gaesorum Raetorum.

Cohors VI Thracum

Cohors VI Thracum is not mentioned on any British diploma. Between AD 80 and 164 they are frequently attested in Germany, Pannonia, Moesia and Dacia. Only two pieces of evidence for the unit have arisen in Britain. The first is the tombstone of one Rufus Sita of the VI Thracum from Wootton, near Gloucester. The tombstone has been dated on formulaic grounds to before AD 61, and so it may be that the unit came to Britain in AD 43 for the invasion. The second evidence is a series of lead seals from Brough-under-Stainmore, which could have come from any other site, since lead seals would move around with messages, the presence of a large number at one site does raise the possibility of the cohort garrisoning the fort at Brough. The first fort at Brough is Flavian, possibly dating to the campaigns of Cerialis in AD 71, and it may be that IV Thracum stayed from 43 until the mid-Flavian era, occupying the fort of Brough, before moving to Germany in time for the diploma of AD 80.

Cohors VII Thracum

The existence of this unit in Britain is confirmed between AD 122 and AD 178 on no less than six diplomas. The only epigraphic evidence we have is three finds of lead seals, one at High Rochester and two at Brough-under-Stainmore, all of which could have come from any site.

Cohors II Vasconum

Little can be confirmed of this unit's British service. The only evidence for their existence comes from diplomas, which place the unit in Britannia between AD 105 and AD 122. The cohort should not be equated with Cohors II Hispana Vasconum, despite the similarity, as the latter is confirmed by diplomas to have been in Mauretania Tingitana during this period. We are told that Galba raised cohorts from the Vascones for service in Germany, and this may well be one of those units (Tacitus: *Histories* IV.33).

Chapter 9

Irregular Units

Cuneus Armatarum
Cuneus Frisiavonum Aballavensium
Cuneus Frisiavonum Vercovensium
Cuneus Frisiavonum Vinoviensium
Equites Catafracariorum
Equites Catafractarii Iuniores
Equites Crispianorum
Equites Dalmatarum (and Branodunensium)
Equites Honoriani Seniores
Equites Scutarii Aureliaci
Equites Stablesianorum Gariannonensium
Equites Syri
Milites Tungrecanorum
Numerus Abulcorum
Numerus Barcariorum
Numerus Barcariorum Tigrisensium
Numerus Concangensium
Numerus Defensorum
Numerus Directorum
Numerus Equitum Sarmatarum
Numerus Equitum Stratonicianorum
Numerus Exploratorum Bremeniensium
Numerus Exploratorum Habitancenses
Numerus Exploratorum
Numerus Fortensium
Numerus Hnaudifridi
Numerus Longovicianorum
Numerus Magnesium
Numerus Maurorum Aurelianorum
Numerus Nerviorum Dictensium
Numerus Pacensium
Numerus Solensium
Numerus Supervenientium Petueriensium
Numerus Syrorum Saggitariorum
Numerus Turnacensium
Numerus Vigilum
Primani Iuniores
Venatores Bannienses
Vexillatio Germanorum Voredis
Vexillatio Marsacorum
Vexillatio Equitum Provinciae Germaniae
Vexillatio Gaesatorum Raetorum
Vexillatio Sueborum Longovicanorum
Victores Iuniores Britanniaciani

Cuneus Armaturum

This occasionally quoted unit is almost certainly an error. This entry in the *Notitia Dignitatum* refers to the fort of Ribchester, where we know an *ala* of Sarmatians to have been based. Any reading of the *Notitia* as 'Cuneus Armatarum' seems likely to be a corruption of 'Cuneus Sarmatarum'.

Cuneus Frisiavonum Aballavensium

We have only one Frisiavonum cohort attested in Britain, yet there is evidence of Frisiavones at three forts in addition to the one that cohort occupied. Cohors I Frisiavonum seems to have been based at Rudchester throughout the third and fourth centuries, and so none of the three *cunei* can refer directly to that unit. A *cuneus* was a smaller, irregular unit, and might be considered a vexillation of a parent cohort, and this might be the case here. However, we are left with several problems. This 'unit of Frisiavones from Aballava' are attested on *RIB* 882 in Papcastle in AD 241. At this time we cannot be certain of the garrison of Aballava (Burgh-by-Sands), though it may well still be Cohors I Nerviorum.

Three possible answers to this conundrum occur. Perhaps a vexillation of I Frisiavonum from Rudchester were posted for a time at Burgh-by-Sands alongside its garrison unit. It might be noted that the terrain on the Solway Firth would bear a resemblance to the Frisiavones' homeland around the mouth of the River Rhine. However, it would have to be a small unit to occupy a fort that already had a full garrison. A second possibility is that I Nerviorum had left Burgh-by-Sands by AD 241, and another Frisiavones' cohort was based there, otherwise unattested in the province. That would certainly account for a small irregular unit from there being present at Papcastle, just 20 miles away. The most attractive answer, though, is that the 'unit of Frisiavones' were actually part of Cohors I Nerviorum. The Nervii's ancient territory lies very close to Frisian lands, and it may be that this represents a group of Frisiavone recruits within the Nervian cohort.

Cuneus Frisiavonum Vercovicensium

As with Cuneus Frisiavonum Aballavensium above, similar issues and solutions present themselves at the fort of Housesteads. Once again, the irregular unit of Frisiavones who raised altars at the fort there (*RIB* 1594 and probably 1593) between AD 222 and 235, could be part of the Rudchester garrison on temporary assignment at Housesteads. We can likely discount the possibility of a purely Frisiavones unit in garrison there at the time, for Cohors I Tungrorum is attested at Housesteads in the Severan era. Once again, though, as with the Aballava group, the Tungri's homelands were as close to Frisian territory as the Nervii's was. We may again be looking at a small group of Frisiavones recruits within the Tungrian cohort there. Interestingly, both altars at Housesteads name the Tuihanti, who seem to have been a subtribe of the Frisii, just as the Frisiavones were.

Cuneus Frisiavonum Vinoviensium

The Frisian *cuneus* attested at Binchester on *RIB* 1036 presents a different case to the other two. The fort's only known garrison were Spaniards, and therefore unlikely to contain Frisiavone recruits. However, we cannot say with any certainty that the Spanish unit, Ala Hispanorum Vettonum, was based at Binchester after AD 198, which opens up an interesting possibility. The wording of the undated altar reads 'EX C FRIS VINOVIE', and it is possible that the C stands not for Cuneus but for cohors. As such, it is quite possible that a second unit of Frisiavones, otherwise unattested, was based at Binchester in the third or fourth century.

Equites Catafractariorum

The *Notitia Dignitatum* tells us of a '*Praefectus equitum catafractariorum, at Morbio.*' Morbium is an unidentified fort, though the name is often linked with Piercebridge. Piercebridge is certainly a fitting possibility, for such investigation as has been achieved has revealed it to have been reconstructed as a cavalry fort circa 260. Morbio has also been linked with Ilkley, although that fort seems to have been designed for a *cohors equitata*. Cataphracts were an eastern invention, a fully armoured rider on an armoured horse, specialized for shock action like medieval knights, and we have no other reference to their presence on the island. However, we do have one people who fit the bill. The Sarmatian cavalry shown on Trajan's Column fit the description of cataphracts perfectly, and bear a striking resemblance to the *draconarius* tombstone at Chester, who was almost certainly a Sarmatian rider from Ribchester. Thus, there is considerable circumstantial evidence to suggest that this unit could be another quingenary unit of Sarmatians of the 5,500 brought to the island by Marcus Aurelius in AD 175, along with those confirmed at Ribchester (Ala Sarmatarum) and those possibly present at Catterick (Numerus Equitum Sarmatarum).

Equites Catafractarii Iuniores

This unit is named as serving the *comes Britanniarum* in the *Notitia Dignitatum*. Given the epithet, it may be that they were a vexillation of the Equites Catafractariorum assigned to the *comes'* army.

Equites Crispianorum

This unit, based in Doncaster according to the *Notitia Dignitatum*, was clearly a cavalry unit. We have no concrete details of the unit other than that, and it was possibly named for Crispus, son of Constantine I. After the condemning and execution of Crispus one might expect a unit to drop the name, though given how little we know of the circumstances surrounding his death, and the loyalty the British forces showed to the Constantinian dynasty, perhaps their cleaving to the name is not so surprising. There is one other intriguing possibility, though. Near modern Zirc, in Hungary and just south of the Danube, was a location known in antiquity as Crispiana. If this cavalry unit took their name from that site, this would make them Pannonian. As such, they may be an otherwise unattested Pannonian unit in Britain, or, possibly, they represent the final form of one of the known units based in Britain. We know the Ala Pannoniorum Tampiana to have gone by AD 130, and we know Ala Pannoniorum Sabiniana were at Halton Chesters, but we have no confirmed details of Cohors II Pannoniorum after Beckfoot in the second century, and none of I Pannoniorum at all other than its presence, so it is possible that either of them became the Equites Crispianorum at Doncaster.

Equites Dalmatarum and Equites Dalmatarum Branodunensium

For these units, both mentioned in the *Notitia Dignitatum*, at Praesidio (Bridlington?) and Branodunum (Brancaster) see Cohors I Delmatarum.

Equites Honoriani Seniores

Another unit assigned to the *comes Britanniarum*, we know nothing of this unit beyond the entry in the *Notitia Dignitatum*. Given their name, they may have been named for Emperor Honorius, who was emperor in the west at the time his cousin, Stilicho, was likely *comes Britanniarum* and rearranging the island's defences (circa AD 395). It is interesting to note that the *Notitia* shows the shield pattern of this unit as two rearing wolves on a red background, which is oddly reminiscent of the tale of the 'Omen of the Wolves', in which Honorius was attacked by two wolves, and though that story is dated to AD 402, the *Notitia* probably postdates the event. This is probably coincidence, but interesting, nonetheless. Interestingly, the *Notitia Dignitatum* names an identical unit serving the *magister equitum Galliarum*, as well as an Equites Honoriani Iuniores, and the two entries probably refer to the same unit. Perhaps it moved from Britain to Gaul during the compilation of the *Notitia*.

Equites Scutarii Aureliaci

Assigned to the army of the *comes Britanniarum* in the *Notitia Dignitatum*, this unit is otherwise unknown. Given its name, it is possible that the unit was brought to the island following Aurelian's reconquest of the breakaway Gallic Empire in AD 274.

Equites Stablesianorum Gariannonensium

Other than their entry in the *Notitia Dignitatum*, which places the unit at Gariannonum (either Burgh Castle or Caister-on-Sea in Norfolk), we have no further information. That they were cavalry is our only detail.

The late Saxon Shore fort of Burgh Castle. (*Author's collection*)

Equites Syri

Attempts to link this unit to any other in Britain are dubious at best, for the only known units that could claim a Syrian origin are boatmen and archers, not cavalry. As with several other later units in Britain, it is possible that the unit was brought to the island following Aurelian's reconquest of the breakaway Gallic Empire in AD 274, given that he had previously been in Syria.

Milites Tungrecanorum

The *Notitia Dignitatum* places this group of Tungrian soldiers at the Saxon Shore fort of Dover (Dubris). Since Cohors I Tungrorum was based at Housesteads at this point, and it seems likely that Cohors II Tungrorum were at Castlesteads based on a lacuna in the text, the only possible unit in Britain that this could refer to is the Ala Tungrorum, though the name would imply that they had changed from a cavalry unit to an infantry one, which is unlikely. Perhaps this garrison represents a vexillation of one of the other surviving units, hence its lesser unit designation, or perhaps an otherwise unattested unit, or even a fresh late Roman garrison recruited from across the Channel in Tungrian lands.

Numerus Abulcorum

Noted in the *Notitia Dignitatum* as based at Anderidos (the shore fort of Pevensey), this is the only evidence we have for the unit in Britain. We do, however, have a little background to the unit. They may have been recruited from the region of Avila, ancient Abula, in Spain, and are first recorded in the Roman military in the mid-fourth century. Zosimus, in his *New History* (book 2), tells us of the Battle of Mursa (AD 351) and the forces fielded by Constantius II: 'Amongst the officers that showed great bravery in this battle and fell in it were Arcadius, commander of the unit called Abulci.' It may be that less than a decade later, when his successor, Julian, sent the general Lupicinus to Britain to counter northern raiders, the Numerus Abulcorum was part of his force, which would then have stayed in garrison on the island.

Numerus Barcariorum

Based on the line N. BARC from the altar *RIB* 601 found at Lancaster, the letters have been expanded to this unit, clearly a force of boatmen, which would be of great use on the west coast between Wales and Cumbria, and on the Lune River which extends inland to reach the fort of Over Burrow. Lancaster is confirmed to have had a phase as a late Roman coastal fortress, which was likely the garrison of Cohors III Nerviorum, but a third-century date has been applied to this altar on textual grounds, and so the numerus might predate that other unit. This numerus cannot be connected with the Barcariorum Tigrisensium of South Shields.

Numerus Barcariorum Tigrisensium

Given as the garrison of South Shields in the *Notitia Dignitatum*, the unit's name gives a great deal of information about them. They were sailors from the River Tigris in the Middle East, which raises a number of questions. The only part of the Tigris that lay within Roman territory or even borders it is the uppermost reaches, east of Syria and into Osroene. That entire stretch of the river was not navigable by more than canoes and small rafts. The river becomes navigable by the smallest and shallowest of vessels around Nineveh, well outside Roman territory, and by the time it is truly boat-navigable, it is in deepest Iraq, the ancient lands owned by Parthia, and later by the Sassanians, Rome's bitterest Eastern enemies. So how did men from there end up as Roman *limitanei* in northern Britain?

Clearly the Tigris boatmen came from outside Roman lands, and can only therefore have been recruited as a result of either a political trade or a military victory. In AD 244, Philip the Arab arranged terms with the Sassanian king, and though it was a deal very favourable to the enemy, it is not impossible that Rome acquired a small unit of boatmen as part of the deal. Equally, in AD 261, the Sassanian king suffered defeats at the hands of Rome and Palmyra, and it is even more likely that boatmen would be given to Rome as part of the deal that followed as Shapur withdrew.

South Shields housed Cohors V Gallorum until at least AD 238–244, and so the numerus of Tigris boatmen must have replaced them in the following century and a half. We may have a clue as to when. Upon concluding his war against independent Palmyra in AD 272–273, Emperor Aurellian then came west to restore to Roman control the breakaway Empire of Gaul and Britain. He will have brought with him forces from the East and left them in garrison in the recovered territory, and this might be the origin of Tigris boatmen in northern Britain, which would date their arrival at South Shields to the mid-270s. Any attempt to link the unit with the Equites Syri is wishful thinking, as that latter unit is cavalry rather than nautical.

Numerus Concangensium

Little can be said about this unit. Numerous stamped tiles have been excavated at the fort of Binchester, all stamped with 'N CON', expanded as Numerus Concangensium. With Chester-le-Street (Concangis) and Binchester being less than 15 miles apart this is perfectly reasonable. Roman military sites tended to be production centres, and Chester-le-Street was clearly a tile manufacturing site. That they should end up at a neighbouring fort is perfectly understandable. What this does not tell us, unfortunately, is anything about the unit other than where they were based and that they made tiles. It may refer to Ala II Asturum, recorded there, although that seems unlikely, especially given that the tiles were used in a bath house from the later third century. The *Notitia Dignitatum* puts a unit of watchmen there: 'Praefectus numeri vigilum at Concangios', and it is perhaps that these two *numeri* are the same unit.

Numerus Defensorum

Noted at Kirkby Thore in the *Notitia Dignitiatum*, this is probably a pseudonym for the unit already in garrison there (see Cohors I Hamiorum).

Numerus Directorum

Based at Brough-under-Stainmore in the *Notitia Dignitatum*, there is some doubt as to the nature of the unit. Hassall offered '*directi*' as a suggestion that Brough operated as some sort of communications hub for the north of England, based on the large number of lead seals from documents found there. Jarrett prefers to read the name as referencing sharpshooters. Either option is equally viable, but if we favour the latter, it might be worth recalling that Cohors I Hamiorum, the only dedicated unit of archers recorded on the island, were based only 12 miles away at Kirkby Thore, and it is possibly that the last garrison of Brough was connected.

Numerus Equitum Sarmatarum

The site of Bainesse, near Catterick on the A1 has produced a fascinating find in *RIB* 2479. A stamped tile found circa 1800 and now lost was noted with the legend BSAR. The B could be a misreading of 'E Q', which would allow the unit here named. The only Sarmatian garrison we have confirmed in Britain was the *ala* at Ribchester and remained there until the *Notitia Dignitatum*'s time, and it is possible that this tile somehow made its way from that fort. There is another possibility, however. As noted in the section on the Ala Sarmatarum, Marcus Aurelius is known to have brought 'fifty-five hundred' Sarmatian cavalry to Britain in AD 175. That is a lot of horsemen, and only 500 will have garrisoned the fort of Ribchester. It is likely therefore that other vexillations of Sarmatians were distributed around the north, used to bolster reduced cavalry units, or perhaps even sent off as full *alae* on their own. No garrison has ever been confirmed for Catterick, though weapons excavated there suggest a cavalry presence at the fort. It is therefore more than possible that Catterick was the base for a second Sarmatian *ala* from the mid-second century. See also Ala Sarmatarum and Equites Catafractariorum.

Numerus Equitum Stratonicianorum

This unit is attested at Brougham on an altar (*RIB* 780) dated to the third century by the style of the inscription. They are otherwise unknown in Britain, and their origin is unknown, but interesting. Several cities named Stratonicea are known, but all are in the region of modern Greece and Turkey. The nearest that any known unit serving in Britain came to any of those cities would be one of the Thracian units, two cohorts and an *ala*, and it is not inconceivable that a small unit of riders from one of those Eastern cities could have been recruited into a Thracian unit. Another possibility is that they are an entirely separate unit of the later Roman army. It is tempting to posit a connection with Emperor Aurelian. In AD 272–273, Aurelian campaigned in the East, reconquering the breakaway Palmyrene Empire, in the region of at least two cities by the name of Stratonicea. Only the following year, Aurelian then came west, doing the same thing with the breakaway Gallic Empire, which included Britain. It is not beyond the bounds of possibility that among the forces he brought from the East was a cavalry unit from Stratonicea, which would then have taken the field in Gaul. When the empire was restored, the unit could then have been settled into garrison in Britain, part of a force loyal to the emperor tasked with maintaining control over the island.

Numerus Exploratorum Bremeniensium

This unit of scouts is recorded at High Rochester on altar *RIB* 1270, which has been dated to the third century. At this time Cohors I Vardullorum was likely the fort's regular garrison, and so the scouts seem to have been an extra irregular force attached to the Vardulli. The presence of a dedicated scout unit in the advance forts of Hadrian's Wall in the years following the Severan withdrawal from Scotland is unsurprising. Since High Rochester was not occupied by the time of the *Notitia Dignitatum*, it is tempting to suggest this is one of the units that forms scout forces elsewhere by the fifth century (see below).

Numerus Exploratorum Habitancenses

Like the High Rochester scouts above, this unit seems to be a post-Severan addition to the regular garrison of the outpost fort of Risingham. They are referenced there two inscriptions (*RIB* 1244 and 1235), with the latter dated convincingly to AD 213. That inscription notes that the *numerus* raised the altar in conjucnction with a unit of Raetian Spearmen and Cohors I Vangionum. Risingham was abandoned by the time of the *Notitia Dignitatum*, and this unit may be one of the scout units the work places elsewhere.

Numerus Exploratorum

The *Notitia Dignitatum* puts an irregular unit of scouts at the fort of Bowes, in the Stainmore Pass. There is some sense to a scout unit based at this fort, given its command of an important pass through the Pennines, wild countryside that could be an excellent habitat for bandits. The unit may be one of those previously based at High Rochester or Risingham, transferred when the outpost forts were abandoned.

Numerus Exploratorum

Another scout unit, and again possibly one of those withdrawn from Risingham or High Rochester, is placed by the *Notitia Dignitatum* at the late shore fort of Portchester (Portus Adurni). It is somewhat unexpected to find a unit of scouts based in one of the most settled regions of Britain, and at a port no less. It is possible they were employed more in the form of dispatch riders between the Saxon Shore forts than as scouts *per se*.

Numerus Fortensium

Based at the Saxon Shore fort of Bradwell-on-Sea, according to the *Notitia Dignitatum*. Why the garrison would be known as the 'strong unit', we can hardly guess, and this is the only information we have on them.

Numerus Hnaudifridi

This unit is attested at Housesteads on the altar *RIB* 1576. It appears to reference an individual by the name of Hnaudifridus rather than any group of people, and as such may be identifiable with the Cuneus Frisiavonum Vercovicensium, or with any other vexillation of troops based at the fort. One thing that can be said with some certainty is that the unit were of Germanic origin, given the three deities the altar was raised to.

Numerus Longovicanorum

This unit, attested at Lanchester in the *Notitia Dignitatum*, is rather vague. Given that its name simply refers to 'the unit from Longovicium', it tells us little. In the mid-third century, Cohors I Lingonum are the otherwise-last-recorded garrison of Lanchester, and so they may be the same unit as this one.

Numerus Magnesium

Similar to the Longovicanorum above, this unit, we are told by inscription *RIB* 1825, occupies the fort of Carvoran. Since that fort's Latin name was Magnis, once again, this *numerus* seems to be called 'the unit from Magnis', and could refer to any unit. If we presume the numerus to be a later occupant, it may be equated with Cohors II Delmatarum, who the *Notitia Dignitatum* places there.

Numerus Maurorum Aurelianorum

The *Notitia Dignitatum* places this unit at Burgh-by-Sands. For the possibility that this references Mauretanian recruiting among other units, see Cohors I Nerviorum. It might, of course, refer to an otherwise-unknown unit. Given how the Severans showed favour to Africa, and Caracalla in particular favoured the Mauri berber officer Macrinus, it is tempting to see the presence of Mauri cavalry here as a result of the Severan campaigns in Scotland, especially given their epithet, when Caracalla's regnal name was Marcus *Aurelius* Antoninus.

Numerus Nerviorum Dictensium

Placed by the *Notitia Dignitatum* at the fort of Dicti, this seems likely to be the remnants of one of the Nervian units that had served on the island for centuries. The location of Dicti has never been identified, and is often believed to be the supposed fort at Whitby. That Dicti is also referenced in the seventh-century Ravenna Cosmography as 'Dixio Lugunduno' may change that. *RIB* 1054 seems to tell us that Lugudunum was the earlier name for the fort of South Shields. As such, 'Dicti' may reference an unidentified fort somewhere in the northeast. Regardless, the unit based there was clearly Nervian. Of the Nervian units we have discussed in this work, VI Nerviorum is attested late at Brough-by-Bainbridge, II Nerviorum were last seen at Whitley Castle in the third century, and I Nerviorum at Burgh-by-Sands with no confirmed date beyond the second century. The Ala Nerviorum and cohorts III and IV all have patchy service histories at best. As such, barring VI Nerviorum, any of the other Nervian units in the province could have ended up as the *numerus* at Dicti.

Numerus Pacensium

Once again, the name of this unit is debated. It could mean something like 'peacemakers', which gives us no more information about the unit. An alternative reading might suggest that the unit was raised in the area of Colonia Civitas Pacensis in what is now southern Portugal. This would allow tentative links to III Bracaraugustanorum, another Lusitanian unit, although the two locales being around 250 miles apart, the connection is tenuous at best. All we can really comfortably say is that they were based at the time of the *Notitia Dignitatum* at the fort of Burrow Walls on the Cumbrian coast.

Numerus Solensium

Another *Notitia Dignitatum* reference, this unit was based at the fort of Old Carlisle. The name may be a reference to the sun, but other than that we know nothing about the numerus.

Numerus Supervenientium Petueriensium

This unit is placed by the *Notitia Dignitatum* at the fort of Derventio. That name has long been linked with Malton, which stands on the River Derwent, though recent years have seen more and more arguments against the connection. The identification of local site names is complicated by the *Antonine Itinerary*, which may or may not have a transcription error in the area, chaning the mileages between forts. Malton may be Delgovicia, while Derventio could be a number of other sites in the area. Of those forts, the one at Stamford Bridge is also on the Derwent and has the strongest case to be Derventio.

The unit's name connects them to the town of Brough-on-Humber (Petuaria), which was an important port town and ferry crossing of the Humber. The unit may have previously been stationed there, just 20 miles or so away, and then been moved to Stamford Bridge later on. Another possibility is suggested by the word '*supervenientium*'. Taken from Supervenio, this suggests that the unit had the duty of supervising the port of Petuaria. Stamford Bridge lies at an important crossroads in the East Riding of Yorkshire, where two critical roads meet. One runs from Petuaria up to the supposed fort at Whitby, which may well have been a central control post for the Yorkshire signal stations. The other runs from the legionary fortress and regional capital of York to the coast at Bridlington, long believed to be the site of the fort of Praesidio given on the *Antonine Itinerary*. Given that name, the fort there must have been important. If, then, this *numerus* controlled the crossroads between the four places, that they would supervise the traffic from Petuaria is tempting.

Numerus Syrorum Saggitariorum

This unit, based on the altar *RIB* 764 entirely relies upon Mommsen's expansion of the line 'N M S S' to be '*Numerus Militarum Syrorum Saggiratiorum*'. It is far from being concrete, and yet to date is our best guess. Based on this reading, the unit can be equated with Cohors I Hamiorum, the only known unit of Syrian archers in Britain.

Numerus Turnacensium

We can say nothing of use about this otherwise-unknown unit other than that the *Notitia Dignitatum* places them at the Saxon Shore fort of Lympne (Portus Lemanis).

Numerus Vigilum

This unit of watchmen, the *Notitia Dignitatum* tells us, was based at Chester-le-Street. They may be the same unit as the Numerus Concangensium described above.

Primani Iuniores

Stated in the *Notitia Dignitatum* as serving the *comes Britanniarum*, the name of this unit makes any kind of identification extremely difficult. The clear suggestion is that they were a sub-unit of an *ala* or cohort with the numeral I in the title.

Venatores Bannienses

This group is named on an altar to Silvanus (*RIB* 1905) found at Birdoswald fort on Hadrian's Wall. Rather than being a separate unit, it is more reasonable to assume that these 'hunters from Banna' were *immunes* (soldiers exempt from fatigues) that formed part of the fort's garrison, so likely either Cohors I Tungrorum or Cohors I Thracum. Hunters are noted as being immunes on the Montana inscription (*CIL* III, 7449).

Vexillatio Germanorum Voredis

The 'German vexillation from Voreda' is mentioned on an altar to the 'mother goddesses from overseas', found at either Old Penrith or Brougham, and now lost (*RIB* 920). Voreda being Old Penrith, it is more likely the stone came from

Brougham, as it seems redundant to name the unit's fort of origin on an altar raised in that fort. See also *RIB* 926 from Old Penrith, an altar raised to all the deities, with some disctintly Germanic names attached. The only recorded garrison of Old Penrith is Cohors II Gallorum, and it may be that they took in a group of German recruits, possibly of the Marsacii (see II Gallorum for details). The simple answer is that this unit seems to have been a visiting vexillation of II Gallorum from Old Penrith.

Vexillatio Marsacorum

This group, mentioned on *RIB* 919 from Old Penrith, was probably part of Cohors II Gallorum (see that unit's listing, and Vexillatio Germanorum Voredis above for more information).

Vexillatio Equitum Provinciae Germaniae

RIB 589, a dedication to Marcus Aurelius and Lucius Verus (AD 163–166) from Ribchester, was raised by a vexillation of cavalry from Germany. Though Ribchester was a cavalry fort, the only two attested garrisons are Asturians from Spain and Sarmatians from across the Danube, neither of whom had a remotely German origin. It is possible that this vexillation was a temporary posting or visiting group from another *ala* or cohort in Britain that *did* have German origins, but the inclusion of the word 'province' is interesting. The word is only rarely found on inscriptions, and even then is usually referencing Britannia. The phrasing suggests that, rather than this group being visitors from another Germanic unit in Britain, they were riders actually visiting from Germania itself. It is interesting to note that in the *RIB* 589 inscription, they are noted as raising the dedication under the governor Sextus Calpurnius Agricola, who is known to have been legate of Germania Superior in AD 158, only five years earlier. It is therefore possible that building work at Ribchester was commanded by Calpurnius Agricola and dedicated by a detachment of riders he had brought with him from Germany.

Vexillatio Gaesatorum Raetorum

Mentioned on an altar to Fortuna at Great Chesters (*RIB* 1724), this unit seems to have been a vexillation of spearmen from Raetia (modern Switzerland and environs). The altar has been dated on stylistic grounds to the third century. In AD 166–169, a Raetian unit appears to have garrisoned Great Chesters, for *RIB* 1737 includes the legend '[...] RAETORU[...]'. It may be preceded by the numeral V, making them Cohors V Raetorum. The change in unit type at the fort suggests that sometime between AD 166 and the third century, Cohors V Raetorum had moved to another location, but left a vexillation in the area. Certainly by AD 225, the fort was in the care of Cohors II Asturum.

Interestingly, *RIB* 1235 from Risingham also mentions Raetian spearmen, alongside scouts and the garrison unit of Cohors I Vangionum, and has been dated to AD 213, possibly around the same time as our vexillation at Great Chesters, and less than 20 miles away. *RIB* 1216 and 1217, also from Risingham, seem to both reference this group, through the letters G R in the inscriptions, with 1216 once again being given a third-century date. *RIB* 2117 from Jedburgh, probably originating at the fort of Cappuck, once again mentions Raetian spearmen, under the command of a tribune named Julius Severinus, who is attested at Risingham. This stone cannot be dated, but may well also belong to the same third-century period. The tombstone of a Raetian woman by the name of Titullinia Pussitta was found at Netherby, again less than 20 miles from Great Chesters.

The earliest evidence we have for Raetians in Britain comes from Vindolanda tablet 892, dated AD 85–90. In it, a decurion called Masculus writes to the commander of Chesterholm, one Iulius Verecundus, and asks for a number of men to be sent to him, including three Raetians, stating that they have been requested by other Raetians, presumably these ones under Masculus' command. A last piece of Raetian evidence comes in the form of *RIB* 576 from Manchester, an altar which references a vexillation of Raetians and Noricans.

Given that the third-century vexillation are spearmen, the Great Chesters garrison a cohort, and the first-century unit under the command of a decurion, we have three very different groups. This could suggest more than one unit at work, though the proximity of most of the sites mentioned does rather suggest only one unit. If an *equitata* cohort of Raetians had been present on the island, perhaps since the initial invasion, or maybe since the Flavian expansions, they could have contained cavalry decurions in the first century, been a garrison cohort in the second century, and supplied vexillations of spearmen in the third. It seems, on balance, that Cohors V Raetorum were a busy and flexible unit.

Vexillatio Sueborum Longovicanorum

This group is named on the altar *RIB* 1074 found at Lanchester fort. The only attested full units at Lanchester are Vardulli and Lingones, both of whose tribal lands are far removed from the Suebi and so this vexillation must be connected with a unit from elsewhere, and presumably a unit otherwise unattested. Though the altar places this vexillation at Lanchester during the reign of Gordian III

(AD 238–244), there might be reason to suggest the presence of a Suebian unit in Britain considerably earlier. During the Marcomannic Wars, Cassius Dio tells us of Suebian captives and those who surrender: 'Some of them were sent on campaigns elsewhere, as were also the captives and deserters who were fit for service.' Also, during that war, a group of Germanics living at Ravenna revolted, and again Cassius Dio tells us 'Marcus [Aurelius] did not again bring any of the barbarians into Italy, but even banished those who had previously come there.'

While there is no other evidence for a Suebian unit in Britain, it should be remembered that military diplomas for the *auxilia* were only in use between the reigns of Claudius and Caracalla, and the latest known diploma for Britain is dated AD 178, possibly before any Suebi unit arrived on the island. As such one might not expect to find evidence other than in the form of inscriptions from forts. It is not hard, then, to see how a unit of Suebi could end up serving in Britain sometime in the AD 170s, a couple of generations before the altar was raised at Lanchester.

Victores Iuniores Britanniciani

Serving the *comes Britanniarum*, according to the *Notitia Dignitatum*, these may be the 'Victores' we are told the general Theodosius brought to Britain to counter the Great Conspiracy in AD 367.

Part III

Appendices

Appendix 1

A Listing of the Forts of Roman Britain and their Garrisons

This list is compiled from the evidence in this text. It includes all known and confirmed units, with dates if they can be estimated or proved, as well as any *numeri* or miscellaneous units that may have formed the garrison at other times, unless they conflict with known occupants (where irregulars may have been assigned alongside a principal garrison, they have not been added).

Ardoch (Alauna)
- Cohors I Hispanorum 79–85

Bar Hill
- Cohors I Baetasiorum 142–165
- Cohors I Hamiorum 142–165

Beckfoot (Bibra)
- Cohors II Pannoniorum 105–124

Benwell (Condercum)
- Ala I Hispanorum Asturum 128–410

Binchester (Vinovia)
- Ala Hispanorum Vettonum c.197–198

Birdoswald (Banna)
- Cohors I Dacorum 128–410

Birrens (Blatobulgium)
- Cohors I Nerviorum 142–156

Bowes (Lavatris)
- Cohors IV Breucorum 128–198
- Cohors I Thracum 198–
- Numerus Exploratorum –410

Bradwell-on-Sea (Othona)
- Numerus Fortensium –410

Brancaster (Branodunum)
- Cohors I Aquitanorum 220–
- Equites Dalmatarum Branodunensium –410

Brecon Gaer (Cicucium)
- Ala Nerviorum c.71–130
- Ala Hispanorum Vettonum first to second century

Bridlington? (Praesidio)
- Equites Dalmatarum –410

Brough-by-Bainbridge (Virodsidium)
- Cohors VI Nerviorum 211–410

Brough-on-Noe (Navio)
- Cohors I Aquitanorum 142–220

Brough-under-Stainmore (Verteris)
- Numerus Directorum –410

Brougham (Brocavum)
- Numerus Equitum Stratonicianorum 274–

Burgh Castle (Gariannum)
- Equites Stablesianorum Gariannonensium –410

Burgh-by-Sands (Aballava)
- Cohors I Nerviorum 156–
- Numerus Maurorum Aurelianorum –410

Burrow Walls (Magis)
- Numerus Pacensium –410

Caer Gai
- Cohors I Nerviorum 77–142

Caernarvon (Segontium)
- Cohors I Sunucorum 78–410

Carlisle (Luguvalium)
- Ala Gallorum Sebosiana 71–79
- Ala Petriana

Carrawburgh (Brocolitia)
- Cohors I Aquitanorum 130
- Cohors I Tungrorum 130–142
- Cohors I Batavorum 198–410

Carvoran (Magnis)
- Cohors I Batavorum 85–128
- Cohors VI Nerviorum 128–136
- Cohors I Hamiorum 136–211
- Cohors II Delmatarum –410

Castle Collen
- Cohors IV Delmatarum 102–122

Castlecary
- Cohors I Tungrorum 142–165
- Cohors I Vardullorum 142–165

Castlesteads
- Cohors IV Gallorum 198–211
- Cohors II Tungrorum 241–410

Castlehill
- Cohors IV Gallorum 142–165

Catterick (Cataractonium)
- Numerus Equitum Sarmatarum –410

Chester-le-Street (Concangis)
- Numerus Concangensium (late third century)
- Numerus Vigilum –410

Chesters (Cilurnum)
- Ala Augusta Ob Virtutem Appellata 128–162
- Cohors I Delmatarum 162–175
- Ala II Asturum 175–410

Chesterholm (Vindolanda)
- Cohors I Tungrorum 85–122
- Cohors VIII Batavorum 102–105
- Cohors II Nerviorum 208–211
- Cohors IV Gallorum 211–410

Cramond
- Cohors II Tungrorum 142–165
- Cohors I Cugernorum 142–165
- Cohors V Gallorum 208–211

Doncaster (Danum)
- Equites Crispianorum –410

Dover (Dubris)
- Milites Tungrecanorum –410

Drumburgh (Concavata)
- Cohors II Lingonum –410

Ebchester (Vindomora)
- Cohors IV Breucorum 211

Elslack (Olenacum)
- Ala I Herculeana –410

Great Chesters (Aesica)
- Cohors V Raetorum (second century)
- Cohors II Asturum 225–410

Halton Chesters (Onnum)
- Ala I Pannoniorum Sabiniana 128–410

Hardknott (Mediobogdum)
- Cohors IV Delmatarum 122–138

High Rochester (Bremenium)
- Cohors I Lingonum 105–211
- Cohors I Vardullorum 211–

Housesteads (Vercovicium)
- Cohors I Hamiorum 122–136
- Cohors I Tungrorum 205–410

Ilkley (Verbeia)
- Cohors II Lingonum 164–

Inveresk
- Ala Gallorum Sebosiana 142–164

Kirkby Thore
- Cohors I Hamiorum 211
- Numerus Defensorum –410

Lancaster (Calunium)
- Ala Augusta Gallorum Proculeiana 98–164
- Ala Gallorum Sebosiana 164–
- Cohors III Nerviorum –410

Lanchester (Longovicium)
- Cohors I Vardullorum 165–211
- Cohors I Lingonum 211–
- Numerus Longovicianorum –410

Llanio (Bremia)
- Cohors II Asturum (Flavian?)

Lympne (Portus Lemanis)
- Numerus Turnacensium –410

Malton? (Derventio)
- Ala Gallorum Picentiana c.84–
- Numerus Supervenientium Petueriensium –410

Manchester (Mamucium)
- Cohors III Bracaraugustanorum 103–160
- Cohors I Frisiavonum 160–211
- Cohors I Baetasiorum 211–212

Maryport (Alauna Carvetiorum)
- Cohors I Aelia Hispanorum 128–138
- Cohors I Delmatarum 138–162
- Cohors I Baetasiorum 162–208

Melandra Castle (Ardotalia)
- Cohors I Frisiavonum 110–140

Moresby (Gabrosentum)
- Cohors II Lingonum 128–164
- Cohors II Thracum 164–

Mumrills (Pexa)
- Ala I Tungrorum 142–164
- Cohors II Thracum 142–164

Netherby (Castra Exploratorum)
- Cohors I Aelia Hispanorum 208–

Newcastle-upon-Tyne (Pons Aelius)
- Cohors I Cugernorum 265–410

Newstead (Trimontium)
- Ala Augusta Vocontiorum 142–164

Old Carlisle (Maglona)
- Ala Augusta Ob Virtutem Appellata 175–242
- Numerus Solensium –410

Old Kilpatrick
- Cohors I Baetasiorum 208–211

Old Penrith (Voreda)
- Cohors II Gallorum 122–235

Papcastle (Deventio)
- Cohors I Vangionum –208

Pevensey (Anderita)
- Numerus Abulcorum –410

Piercebridge? (Morbio)
- Equites Catafractariorum –410

Portchester (Portus Adurni)
- Numerus Exploratorum –410

Ravenglass? (Glannoventa)
- Cohors I Morinorum –410

Reculver (Regulbium)
- Cohors I Baetasiorum 212–410

Ribchester (Bremetenacum)
- Ala II Asturum 122–175
- Ala Sarmatarum 175–410

Richborough (Portus Rutupiae)
- Legio II Augusta c.273–410

Risingham (Habitancum)
- Cohors I Vangionum 208–

Rough Castle
- Cohors VI Nerviorum 142–165

Rudchester (Vindobala)
- Cohors I Frisiavonum 211–410

Slack (Cambodunum)
- Cohors IV Breucorum 98–128

South Shields (Arbeia)
- Cohors V Gallorum 211–274
- Numerus Barcariorum Tigrisensium –410

Stanwix (Uxellodunum)
- Ala Petriana 160–410

Templebrough
- Cohors IV Gallorum 117–142

Wallsend (Segedunum)
- Cohors II Nerviorum 128–208
- Cohors IV Lingonum 208–410

Whitby (Dictium)
- Numerus Nerviorum Dictensium –410

Whitley Castle (Epiacum)
- Cohors II Nerviorum 211–410

Appendix 2

A Timeline for Roman Britain

43	Invasion of Britain under Aulus Plautius
46	Thracia added to the empire as a province
60	Invasion of Anglesey by Suetonius Paulinus
61	Boudiccan Revolt
69	Year of the Four Emperors. Revolt of Civilis in Germany
71	Cerialis founds York and campaigns against Brigantes
77	Agricola made governor of Britain. Conquest of Anglesey
83	Battle of Mons Graupius. Domitian's Chatti War
86	Domitian's Dacian War begins
87	Scotland abandoned. Army withdrawn south
88	Domitian's Dacian War concludes
96	Domitian's death. Trajan in Germany
98	Accession of Trajan. Ager Nerviorum diploma
101–102	Trajan's First Dacian War
103	Malpas Diploma
105	Sydenham Diploma, Trajan's Second Dacian War begins
106	Trajan's Second Dacian War ends
119	Unrest in Britain
122	Construction begins on Hadrian's Wall. Brigetio Diploma
124	Stannington Diploma
126	Walcot Diploma
128	Hadrian's Wall is completed
135	Wroxeter Diploma
141	Lollius Urbicus campaigns in lowland Scotland
142	Construction begins on Antonine Wall
145	Chesters Diploma
146	Vindolanda Diploma
158	Ravenglass Diploma
159	Colchester Diploma
162	Antonine Wall abandoned
166	Marcomannic Wars begin

175	5,500 Sarmatians sent to Britain by Marcus Aurelius
178	Bulgaria Diploma
182	Revolt of the Brigantes
193	Hadrian's Wall damaged in raids. Severan dynasty begins
195	Clodius Albinus takes men from Britain to the continent
209	Septimius Severus campaigns in Scotland. Hadrian's Wall reworked
211	Caracalla made emperor in York. Severan withdrawal from Scotland
212	Caracalla's edict of universal citizenship
235	Severan dynasty ends
259	Cessation of the 'Gallic Empire' under Postumus
272	Aurelian campaigns against 'Palmyrene Empire'
273	Aurelian completes recovery of the East
274	Gallic Empire reconquered by Aurelian
286	Carausius forms breakaway empire
289	Maximian campaigns against Carausius
293	Work begins on Saxon Shore sites
296	Breakaway empire reconquered by Constantius Chlorus
305	Constantine joins Constantius in Britain. War in Scotland
306	Constantine made emperor in York
314	Raids by Pictii, Scotii and Caledonii
343	Visit of emperor Constans, campaigning against the Picts
351	Battle of Mursa
367	The 'Great Conspiracy' in Northern Britain
368	Flavius Theodosius restores control
383	Magnus Maximus proclaimed emperor in Britain
393	Accession of Honorius
396	Rise of Stilicho
398	Stilicho in Britain fighting the picts
400	Stilicho repairs Hadrian's Wall
402	Battle of Pollentia. Stilicho withdraws troops from Britain
406	Revolt of Marcus in Britain
407	Constantine III made emperor in Britain
410	Honorius withdraws Roman support for Britannia

Appendix 3

Gods Worshipped on Altars in Roman Britain, Grouped by Ethnic Origin

	Roman/Greek	Local/Celtic	German	Eastern
1	Aesculapius	Abandinus	Alaisiagae	Astarte
2	Aeternae	Ancasta	Beda	Hammia
3	Anicetus	Antenociticus	Boudihillia	Herakles
4	Bellona	Arecurius	Fimmilena	Syria
5	Bona Dea	Arnomectae	Friagabis	
6	Bonus Eventus	Belatucader	Harimella	
7	Britannia	Bellinus	Maponus	
8	Ceres	Brigantia	Vanauns	
9	Concordia	Camulos	Vernostonus	
10	Diana	Cocidius	Viradecthis	
11	Discipline	Coventina		
12	Fata Bona	Digenis		
13	Fontes	Epona		
14	Fortuna	Ialonus Contrebi		
15	Genius	Latus		
16	Hercules	Matunus		
17	Hygaeia	Mogons		
18	Isis	Nodens		
19	Juno	Ocelus		
20	Jupiter	Ricagumbeda		
21	Mars	Sattada		
22	Matres	Sulis		
23	Mercury	Toutatis		
24	Minerva	Verbeia		
25	Mithras	Veteris		
26	Nemesis	Vinotonus		
27	Neptune	Viridius		
28	Nymphae	Arciarcon		
29	Oceanus	Garmangabi		

	Roman/Greek	Local/Celtic	German	Eastern
30	Onirus			
31	Pacis			
32	Panthea			
33	Parcae			
34	Priapus			
35	Regina			
36	Roma			
37	Romulus			
38	Saturn			
39	Serapis			
40	Silvanus			
41	Sol			
42	Somnus			
43	Soter/Saviours			
44	Tethys			
45	Vacuna			
46	Victoria			
47	Virtue			
48	Volcanus			

Bibliography

Epigraphic Sources

Annee Epigraphique – via JSTOR
Britannia – https://www.cambridge.org/core/journals/britannia
Corpus Inscriptionum Latinarum (*CIL*) – https://cil.bbaw.de/en/
Epigraphik-Datenbank – Clauss/Slaby (*EDCS*) – https://db.edcs.eu/epigr/epi.php?s_sprache=en
Journal of Roman Studies (*JRS*) – https://www.cambridge.org/core/journals/journal-of-roman-studies
Roman Inscriptions of Britain (*RIB*) – https://romaninscriptionsofbritain.org/

Ancient Sources

Ammianus Marcellinus, *Rerum Gestarum* (Loeb Edition, 1935)
Anon., *Historia Augusta* (Loeb Edition, 1960)
Anon., *Notitia Dignitatum* (Cambridge University Press, 2019)
Anon., *Res Gestae of Augustus* (Loeb Edition, 1924)
Arrian, T*actica* (B.G. Teubneri, 1885)
Cassius Dio, *Historia Romana* (Loeb Edition, 1917)
Cornelius Tacitus, *Agricola & Germania* (Mattingly and Handford, 1948)
Cornelius Tacitus, *Annals & Histories* (Loeb Edition, 1937)
Diodorus Siculus, *Bibliotheca Historica* (Via Bill Thayer's Lacus Curtius website)
Gildas, *De Excidio Britanniae* (J.A. Giles, 1841)
Herodian, *Roman History* (Berkeley, 1961)
Hyginus, *De munitionibus castrorum* (Campbell, 2018)
Justinian, *Digest* (Cincinatti, 1932)
Pausanias, *Description of Greece* (Heinemann, 1918)
Polybius, *Histories* (Loeb Edition, 1923)
Strabo, *Geography* (Loeb Edition, 1932)
Suetonius, *Lives of the Twelve Caesars* (Loeb Edition, 1914)
Titus Livius (Livy), *Ab Urbe Condita* (Heinemann, 1924)
Varro, *De Lingua Latina Libri* (Loeb edition, 1938)
Vegetius, *De Re Militari* (Clark, 1767)
Zosimus, *Historia Nova* (Green and Chaplin, 1814)

Main Bibliography

Birley, Anthony, *Garrison Life at Vindolanda* (Tempus, 2002)
Birley, Eric, *Pannonians in Roman Britain* (Dr Rudolf Habelt Ltd, 1988)
Bishop, M.C., *The Crosby Garrett Helmet* (Armatura Press, 2013)
Bishop, M.C., 'The Military Fabrica and the production of arms in the early Principate', in *The production and distribution of Roman military equipment: proceedings of the Second Roman Military Equipment Research Seminar,* BAR S275 (1985)

Bishop, M.C. and & J.C.N. Coulston, *Roman Military Equipment* (Oxbow, 2006)
Bowman, A.K., J.D. Thomas and R. Tomlin, 'The Vindolanda Writing-Tablets', *Britannia* 42 (2011)
Breeze, David J. & Brian Dobson, *Hadrian's Wall* (Penguin, 2000)
Campbell, Duncan B., 'Evidence for Medics among the *Milites*', *Ancient Warfare Magazine* VI.4 (2012)
Campbell, Duncan B., 'The Later Roman Army', *Britannia* 30 (1999)
Cheesman, George, *The Auxilia of the Roman Imperial Army* (Clarendon Press, 1914)
D'Amato, Raffaele, *Decorated Roman Armour* (Frontline, 2017)
Devijver, Hubert, *Prosopographia Militiarum Equestrium Quae Fuerunt Ab Augusto Ad Gallienum* (Cornell, 1987)
Dixon, Karen and Pat Southern, *The Roman Cavalry* (Routledge, 2016)
DuBois, Michael S., *Auxillae* (Lulu, 2015)
Erdkamp, Paul (ed.), *A Companion to the Roman Army* (Wiley-Blackwell, 2007)
Gawronski, Radoslaw Andrzej, 'Roman horsemen against Germanic tribes', *Archaeologica Hereditas* (2018)
Goldsworthy, Adrian, *The Complete Roman Army* (Thames & Hudson, 2003)
Goldsworthy, Adrian, *The Roman Army at War* (Oxford University Press, 1998)
Haynes, Ian, *Blood of the Provinces* (Oxford University Press, 2017)
Holder, Paul, 'Auxiliary Deployment in the Reign of Hadrian' in *Documenting the Roman Army* (University of London Press, 2003)
Holder Paul, *A Roman Military Diploma From Ravenglass, Cumbria* (University of Manchester, 1997)
Jarrett, Michael, 'Non-Legionary Troops in Roman Britain: Part One', *Britannia* 25 (1994)
Jarrett, Michael, 'Thracian Units in the Roman Army', *Israel Exploration Journal* (1969)
Keppie, Lawrence, *The Making of the Roman Army* (Routledge, 2015)
Le Bohec, Yann, *The Roman Imperial Army* (Routledge, 2000)
Narloch, Krzysztof, 'The Cold Face of Battle', *Archaologisches Korrespondenzblatt* (2012)
Onofrei, Cosmin, 'A Military Diploma Regarding Dacia Porolissensis', *Ephemeris Napocensis* XXIII (2013)
Oorthuijs, Jasper, 'Marines and Mariners in the Roman Imperial Fleets' (Open access)
Pearson, *The Roman Shore Forts* (History Press, 2002)
Peddie, John, *Conquest: The Roman Invasion of Britain* (Sutton, 1997)
Peddie, John, *The Roman War Machine* (Sutton, 1996)
Porena, Pierfrancesco, *Smetius, Mommsen E L'epitafio di un Anonimo Cavaliere della Seconda Meta del III Secolo* (Dr Rudolf Habelt Ltd, 2018)
Roxan, Margaret, *Roman Military Diplomas 1954–1977* (UCL, 2017)
Roxan, Margaret, *Roman Military Diplomas 1985–1993* (Institute of Archaeology, 1994)
Scott, Andrew, *Emperors and Usurpers* (Oxford University Press, 2018)
Spaul, John E.H. and Conrad Cichorius, *Ala 2: The Auxiliary Cavalry Units of the Pre-Diocletianic Imperial Roman Army*' (Nectoreca Press, 1994)
Spaul, John E. H, *Cohors 2*: *The Evidence for and a Short History of the Auxiliary Infantry Units of the Imperial Roman Army* (Archaeopress, 2000)
Speidel, Michael, 'The Pay of the Auxilia', *Journal of Roman Studies* (1973)
Speidel, Michael, 'The tactics of light cavalry in Mauretania', *Antiquites Africaines* (1993)
Sumner, Graham, 'Roman Soldiers: Fashion icons or military elite?', in De Moor, Fluck & Linscheid (eds.), *Egypt as a textile hub. Textile interrelationships in the 1st millennium* AD – *Proceedings of the 10th conference of the research group 'Textiles from the Nile Valley' Antwerp, 27– 29 October 2017* (Lannoo, 2019)

Thomas, M.D., *Lorica Segmentata Vol. II – A Catalogue of Finds* (Armatura Press, 2003)
Todd, Malcolm (ed.), *A Companion to Roman Britain* (Wiley-Blackwell, 2007)
Turney, Simon, *Agricola* (Amberley, 2022)
Ward, John, *Romano-British Buildings and Earthworks* (Methuen, 1911)
Webster, Graham, *The Possible Effects on Britain of the Fall of Magnentius* (Oxford University Press, 1983)
Webster, Graham, *The Roman Imperial Army* (Black, 1974)
Webster, Graham, *The Roman Invasion of Britain* (Routledge, 2016)
Wilson, Roger, *Roman Forts* (Bergstrom & Boyle, 1980)
Zahariade, Mihail, *The Thracians in the Roman Imperial Army I – Auxilia* (Mega, 2009)

Index